Pathogen removal in aerobic granul systems

Mary Luz Barrios Hernández

Cover picture: C. Boersma

Pathogen removal in aerobic granular sludge treatment systems

DISSERTATION

Submitted in fulfillment of the requirements of
the Board for Doctorates of Delft University of Technology
and
of the Academic Board of the IHE Delft
Institute for Water Education
for
the Degree of DOCTOR
to be defended in public on
Thursday, 30 September 2021, at 12:30 hours
in Delft, the Netherlands

by

Mary Luz BARRIOS HERNÁNDEZ

Master of Science in Urban Water and Sanitation, IHE Delft Institute of
Water Education, the Netherlands
Environmental Engineer, Instituto Tecnológico de Costa Rica, Costa Rica
born in San José, Costa Rica

This dissertation has been approved by the

Promotors: Prof.dr. D. Brdjanovic and Prof.dr.ir. M.C.M. van Loosdrecht
Copromotor: Dr.ir. C.M. Hooijmans

Composition of the Doctoral Committee:

Rector Magnificus TU Delft	Chairman
Rector IHE Delft	Vice-Chairman
Prof.dr. D. Brdjanovic	IHE Delft / TU Delft, promotor
Prof.dr.ir. M.C.M. van Loosdrecht	TU Delft, promotor
Dr.ir. C.M. Hooijmans	IHE Delft, copromotor

Independent members:

Prof.dr.ir. M.K. de Kreuk	TU Delft
Prof.dr. R.H.R. Costa	Universidade Federal de Santa Catarina, Brazil
Prof. G. Chen	Hong Kong University of Science and Technology, China
Prof.dr.ir. E.I.P. Volcke	Ghent University, Belgium
Prof.dr.ir. J.A. Roelvink	IHE Delft, TU Delft, reserve member

This research was conducted under the auspices of the Graduate School for Socio-Economic and Natural Sciences of the Environment (SENSE)

This thesis was accomplished thanks to the financial support of Instituto Tecnológico de Costa Rica.

CRC Press/Balkema is an imprint of the Taylor & Francis Group, an informa business
© 2021, Mary Luz Barrios Hernández

Published by:
CRC Press/Balkema
enquiries@taylorandfrancis.com
www.crcpress.com – www.taylorandfrancis.com
ISBN 978-1-032-13948-7

To my loved ones Arne, Santiago, Federico

ACKNOWLEDGEMENTS

I can do all things through Him (Jesus), who strengthens me.

Philippians 4:13

First and above all, I would like to thank my Lord Almighty, to whom I owe my very existence, for providing me everything in life. This journey could have not been done at all without him.

Of course, there are wonderful people to whom I am very grateful and I would like to acknowledge. I would like to start with my supervisory team. My promotors Dr **Mark** C.M van Loosdrecht (TU-Delft) and Dr **Damir** Brdjanovic (IHE Delft), thanks for allowing me to pursue a PhD; your support and guidance was vital throughout this journey. I would like to give a special thanks to my co-promotor and daily supervisor, Dr Christine M. Hooijmans. Apart from your expertise and significant contribution to this research topic, your diligent supervision, constant guidance, professional and even emotional support was essential for this successful completion. You know in detail the ups and downs of this journey. Thank you for your support **Tineke**! I would like to equally thank my most committed mentor Dr **Hector** Garcia Hernandez. Your mind-sets, expertise, and contribution to this research topic made it enjoyable.

Next, I would like to acknowledge the Instituto Tecnológico de Costa Rica (TEC) for the financial support to pursue this PhD. This journey could not have been possible without my former colleagues' trust (School of Chemistry), Dr. **Luis** Romero from the Environmental Protection Research Center (CIPA), the former and the current authorities of the TEC. I am very grateful for the opportunity to pursue my MSc and PhD studies during these last six years via the World Bank Research Grant and the TEC fellowship department financial support. Thank you, **Floria** Roa, **Marianela** Rojas, for being my financial guarantors.

I would like to show my gratitude to external collaborators from the RoyalHaskoningDHV, Amersfoort, The Netherlands, **Mario** Pronk, **Andreas** Giesen, **Arne** Boersma, **Edward** van Dijk; a special thanks to the operators of the Dutch water boards **Margriethus** van Herk (Waterschap Vechtstromen) **Pascal** Kamminga (Waterschap Noorderzijlvest) for your valuable assistance during the pathogen sampling campaigns.

I also would like to express my gratitude to all the students that contributed to this research - **Carolina, Claribel, Mylene, Karen, Valentina, Maria Clara, Zargham, Ines, Thomas, Laura**. Your trust and motivation kept me going. I learned a lot from you all.

Recognition must be given to all the laboratory staff at IHE Delft, **Fred, Berend, Peter, Lyzette, Ferdi, Frank, Zina**. Your always positive collaboration made my going (up and downs stairs) easier. Thank you for your trust, support and provisions in this project. Berend Lolkema, without your friendship, prompt solutions, imagination, creativity and your always determination to help, we (my students and I) could not have gone through all our professional ups and downs. A genuine thank you for all our chats. Peter Heerings, *E. coli* and I could not survive without you at IHE lab. Thank you for your support in the microbiology lab and for your parental caring.

I also would like to thank my colleagues from TU Delft, **Mona**, **Bruno** for the tailor-made training and determination to share your experience with me. A unique appreciation is given to Dr. **Danny** de Graaff. Your coaching skills in the first year of my PhD were noteworthy in this research project.

Carolien Boersma, "*A picture is worth a thousand words.*" (Brisbane, A). Thank you for making that possible in this book.

The list of friends and acquaintances on the road is extensive and may be incomplete. There have been many years of seeing people coming and going as passengers, but some stayed on the trip a bit longer. **Iosif**, my friend, you got off the bus at an earlier station, but thanks to your friendship, I did not feel alone at IHE during lunchtimes; you are a great reliable person. **Yuli**, I found a friend when I met you. Thank you for the Indonesian lunches we shared and the times we laughed together. Dr. **Ahmed**, my officemate and adopted brother, thank you for your professional and wholehearted support. Your commitment to research has always been my driver. **Mohanad A.**, I am proud of you; your patience and perseverance are unique. Thank you for the endless chats and good ideas. Ex-office neighbours, **Neyler**, **Alida** thanks for the funny talks, advice and friendship. **Mohaned** Sousi, **Muthasen**, **Taha**, **Angelica**, **Mauricio**, **Juanca**, you were part of the journey at different time's life. It was pleasant sharing experiences and social gatherings with some of you guys.

To my most authentic friends, **Cris**, **Nela**, **Clari**, an immense thanks for believing in me, your companion, visits the Netherlands and unconditional support were significant factors to accomplish this journey. You are sisters for me!

Finally, and most importantly, my sincere appreciations to my family. **Arne**, thanks for becoming such an exceptional person in my life and being unconditionally supportive during the last years of my journey. **Santiago**, **Federico**, my lovely sons. You always made me forget the struggles of a PhD life; with you guys, this journey was an adventure. Thanks for keeping me motivated while doing this and for being such supportive boys. **Mami**, **Papi**, **Tita**, **Karol**, **Leroy**, **Cali**, **Sahian**, **Berny**, y todo el resto de la gigantesca familia, muchas gracias por estar ahí siempre, sin importar las condiciones. Su amor a distancia, video llamadas a mitad de la noche y fiestas en mi nombre me hicieron siempre sentir como si estuviera en Costa Rica.

SUMMARY

Pathogens removal from wastewater is crucial for preserving public health. Advanced wastewater treatment plants such as conventional activated sludge (CAS) systems are typically designed for organic matter and nutrient removal, pathogenic organisms are only removed to a limited extend. Attachment to particles followed by protozoa predation are identified as the mechanisms behind pathogen removal in CAS systems. Attached organisms leave the system via the wastage. The aerobic granular sludge (AGS) system is a relatively new and effective wastewater treatment technology that can handle physicochemical constituents' removal simultaneously and in a single process stage. To anticipate if and how pathogen removal in the AGS system takes place, the process needs to be evaluated; in full-scale applications to measure the removal under real life conditions as well as in laboratory-scale reactors to separately assess the different removal mechanisms. This PhD thesis focusses on the mechanisms behind the pathogen removal and degradation processes in AGS systems.

As a first step, pathogen removal was assessed under real life conditions in two full-scale AGS wastewater treatment systems located in the Netherlands, and compared with CAS process technologies (Chapter 2). The AGS and the CAS systems studied received the same wastewater. The removal efficiencies of different faecal indicators are described; among them faecal bacteria (*E. coli*, *Enterococci*, thermotolerant coliforms) and F-specific RNA bacteriophages which are broadly used as viral indicators. The AGS processes showed to be as efficient as the CAS systems in removing the bacterial and viral indicators when treating the same influent. As settling of sludge in the AGS system does not take place in a separate tank, it was not clear what the mechanisms behind the pathogen removal were. Therefore, further studies were carried out in the laboratory and are described in Chapter 3 and 4, showing, among others, the advantage that the granular morphology can give to predators existing in the reactors, and how that enhances the removal of pathogens.

Once the faecal indicator's fate in full-scale applications was elucidated, the mechanisms behind the removal were studied and described. The fate of *E. coli* bacteria and MS2 bacteriophages were explored using laboratory-scale reactors fed with acetate as a carbon source. Reactor conditions simulated full-scale AGS treatment plants: one-hour anaerobic feeding, followed by an extended aeration period and short settling time. Their enumeration after every operational condition, made their fate in the system clear. For both microorganisms, attachment to the granular surface took immediately place after feeding, saturating granules and reaching an equilibrium with the liquid bulk during steady state. The suspended indicators were then exposed to predators during the aeration; and protozoa predation occurred. The *E. coli* bacteria removal was higher than the MS2 bacteriophages removal, and correlated with the protozoa abundance determined using

18S RNA gene sequencing analysis. *E. coli* bacteria, labelled with dsGreen gel staining solution, showed that the bacteria ended up in the ciliates protozoa's vacuoles. The sequencing analysis confirmed the occurrence of those attached protozoa up to genus-level, i.e., *Epistylis, Vorticella,* and *Pseudovorticella.* The analysis also revealed a high abundance of free-living organisms (i.e., *Rhogostoma, Telotrochidium*) which could not be detected by conventional microscopy.

The effect of particulate material on the granular formation and higher-level organisms' development was also investigated and is described in Chapter 4. Synthetic wastewater and faecal sludge (4 % v/v of the total influent) was co-treated in a sequencing batch, laboratory-scale reactor. An anaerobic stand-by enhanced the performance of the system. Overall, the particulate matter affected the granular bed and size of the granules. Part of the granular bed turned flocculent, an accumulation of solids was observed, and granules were reduced in size but kept compact. Attention was given to the protozoa bloom occurring after the faecal sludge addition. They proliferated in the system, attached to the suspended solids and granular surfaces, and contributed to the solids removal. The protozoa did not impact the granules negatively, and they seemed not to be affected by the FS recipes composition.

Finally, to identify the protozoa involved in the pathogen degradation processes in real life applications, sludge samples from full-scale WWTPs were examined using *18S RNA gene* sequencing analysis. Diversity, richness, evenness, abundance and different eukaryotic taxa-level classifications are described in Chapter 5. The analysis was done for raw wastewater, aerobic granules (mixed, large and small fractions), and activated sludge. Operational Taxonomic Units (OTUs) revealed a higher microbial diversity of the raw wastewater compared to the sludge samples. The phylogenetic affiliation (phylum-level) showed organisms that belong to protist, fungi, metazoan, archaea or algae domains. Regardless of the location and WWTPs configuration, the protist community was the dominant group in all analysed samples followed by Ascomiycota (Fungi). To the species-level, significantly abundant structures showed differences between granular and activated sludge but no strong tendencies were measured.

This dissertation identified significant mechanisms playing a role in pathogen removal in AGS systems. The microbial eukaryotic community appeared to be very diverse. Post-disinfection remains essential in case of reuse of treated effluent.

SAMENVATTING

Het verwijderen van pathogenen uit afvalwater is cruciaal voor het behoud van de volksgezondheid. Geavanceerde afvalwaterzuiveringsinstallaties (AWZI's) zoals conventionele systemen met actief slib (CAS) zijn specifiek ontworpen voor het verwijderen van organisch materiaal en nutriënten, pathogene organismen worden slechts in beperkte mate verwijderd. In CAS-systemen worden pathogenen verwijderd door aanhechting aan en bezinking van slibdeeltjes en door predatie door protozoa. Het aëroob korrelslibsysteem (AGS) is een relatief nieuwe en effectieve afvalwaterzuiveringstechnologie waarin de verwijdering van fysisch-chemische bestanddelen tegelijkertijd en in één processtap plaatsvindt. Om te voorspellen of en hoe pathogeenverwijdering in het AGS-systeem gebeurd, moet het proces worden geëvalueerd; in systemen op ware grootte om de verwijdering onder reële omstandigheden te bepalen, alsook in reactoren op laboratoriumschaal om de verschillende verwijderingsmechanismen afzonderlijk te kunnen beoordelen. Dit proefschrift richt zich op de mechanismen achter de pathogeenverwijderings- en afbraakprocessen in AGS-systemen.

Als eerste stap werd de verwijdering van pathogenen bepaald onder reële omstandigheden in twee grootschalige AGS-AWZI's in Nederland, en vergeleken met CAS-procestechnologieën (Hoofdstuk 2). De onderzochte AGS- en CAS-systemen kregen hetzelfde afvalwater als influent. De verwijderingsrendementen van verschillende fecale indicatoren worden beschreven; waaronder fecale bacteriën (*E. coli*, enterokokken, thermotolerante coliformen) en F-specifieke RNA-bacteriofagen die algemeen worden gebruikt als virale indicatoren. De AGS-processen bleken even efficiënt te zijn als de CAS-systemen in het verwijderen van de bacteriële en virale indicatoren bij de behandeling van hetzelfde influent. Omdat bezinking van slib in het AGS-systeem niet in een separate slib-water afscheider plaatsvindt, was het niet duidelijk wat de mechanismen achter de verwijdering van ziekteverwekkers waren. Daarom werden verdere studies uitgevoerd in het laboratorium en dezen zijn beschreven in Hoofdstuk 3 en 4, die onder andere het voordeel laten zien dat de granulaire morfologie kan geven aan predatoren die in de reactoren aanwezig zijn, en hoe dat de verwijdering van pathogenen bevordert.

Nadat het lot van de fecale indicatoren in grootschalige installaties was opgehelderd, werden de mechanismen achter de verwijdering bestudeerd en beschreven. Het lot van *E. coli* bacteriën en MS2-bacteriofagen werd onderzocht met behulp van reactoren op laboratoriumschaal die werden gevoed met acetaat als koolstofbron. Reactorcondities simuleerden AGS-AWZI's: anaërobe voeding van één uur, gevolgd door een langere beluchtingsperiode en een korte bezinkingstijd. Onmiddellijk na het toedienen van de voeding vond aanhechting aan en verzadiging van het oppervlak van het korrelslib plaats, voor zowel de *E. coli* bacteriën als de MS2-bacteriofagen, waardoor een evenwicht werd bereikt met de concentratie in het medium tijdens stationaire toestand. De indicatoren in het medium werden vervolgens tijdens de beluchting blootgesteld aan predatoren; en protozoa-predatie trad vervolgens op. De *E. coli* bacteriën werden in grotere aantallen verwijderd dan de MS2-bacteriofagen en correleerden met de hoeveelheid protozoa bepaald met behulp van 18S rRNA sequentieanalyse. Door middel van labelling met een dsGreen gelkleuringsoplossing van de *E. coli* bacteriën werd aangetoond dat de bacteriën in de vacuolen van de ciliaten terechtkwamen. De sequentieanalyse bevestigde het voorkomen van de aangehechte protozoa tot op genusniveau, d.w.z. *Epistylis, Vorticella* en *Pseudovorticella*. De analyse bracht ook een grote hoeveelheid vrijlevende organismen aan het licht (d.w.z. *Rhogostoma, Telotrochidium*) die niet konden worden gedetecteerd met conventionele microscopie.

Het effect van zwevende deeltjes op de korrelvorming en de ontwikkeling van hogere organismen werd ook onderzocht en is beschreven in Hoofdstuk 4. Synthetisch afvalwater en fecaal slib (4% v/v van het totale influent) werden samen behandeld in een sequencing batch reactor op laboratoriumschaal. Een anaërobe stand-by periode verbeterde de prestaties van het systeem. Al met al had het niet-opgeloste substraat invloed op het korrelbed en de grootte van de korrels. Een deel van het korrelbed werd vlokkig, er werd een opeenhoping van vaste stoffen waargenomen en de korrels werden kleiner en compact. Ook trad er een bloei van de protozoa op na de toevoeging van fecaal slib. Ze verspreidden zich in het systeem, hechtten zich aan de zwevende deeltjes en korrelslib en droegen bij aan de verwijdering van de zwevende deeltjes. De protozoa hadden geen negatieve invloed op de korrels en ze leken niet te worden beïnvloed door de samenstelling van het fecale slib.

Om de protozoa te identificeren die betrokken zijn bij de afbraakprocessen van pathogenen in de praktijk, werden slibmonsters van AWZI's onderzocht met behulp van 18S rRNA sequentieanalyse. De diversiteit, rijkdom, uniformiteit en classificatie in taxonomische groepen van de eukaryoten is beschreven in Hoofdstuk 5. De analyse werd gedaan voor influent, aëroob korrelslib (gemengde, grote en kleine fracties) en actiefslib. De operationele taxonomische eenheden lieten een grotere microbiële diversiteit van het influent zien in vergelijking met de slibmonsters. De fylogenetische verwantschap liet zien dat de organismen behoren tot de protisten, schimmels, metazoa, archaea en algen. Ongeacht de locatie en de configuratie van de AWZI's, vormden de protisten de dominante groep in alle geanalyseerde monsters, gevolgd door Ascomiycota (schimmels). Op soortniveau waren er verschillen tussen granulair en actief slib, maar er werden geen sterke tendensen gemeten.

Het onderzoek beschreven in dit proefschrift identificeerde significante mechanismen die een rol spelen bij het verwijderen van pathogenen in AGS-systemen. De microbiële eukaryotische gemeenschap bleek zeer divers te zijn. Desinfectie blijft essentieel bij hergebruik van behandeld afvalwater.

CONTENTS

1

INTRODUCTION

This chapter provides a short overview of the development and characteristics of the aerobic granular sludge (AGS) wastewater treatment process. Furthermore, an introduction to the emerging problem of pathogens in the wastewater effluent is given, showing the importance to understand the pathogen removal processes and efficiency of AGS systems.

1.1 THEORETICAL BACKGROUND

An overview of the development and characteristics of the wastewater treatment processes studied in this dissertation is provided below. Attention is given to conventional activated sludge system as a leading technology in comparison with the AGS process.

1.1.1 AGS technology development

Wastewater treatment technologies are based on biochemical, organic matter and nutrients conversions [1, 2]. An old but still applied technology is the CAS system [3]. A CAS system's basic configuration consists of an aeration tank where biological conversion occurs, followed by sludge separation using settling tanks (clarifiers) [4]. The sludge age is controlled by recirculating a large fraction of the sludge from the settling tank's underflow to the aeration tank. Advanced configurations have been implemented to enhance the phosphorus uptake and the nitrification/denitrification capacity [5]. As an example, Figure 1.1 shows the University of Cape Town activated sludge configuration [6]. Anaerobic and anoxic compartments are added, where naturally selected micro-organisms carry out organic matter and nutrient conversions.

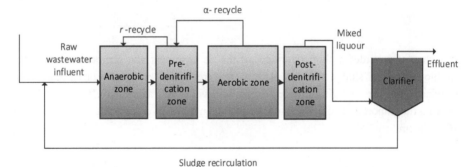

Figure 1.1. University of Cape Town (UCT) configuration of an activated sludge system [5, 7]

In the nineties, a new technology based on granular sludge demonstrated to be optimal for simultaneous organic matter and nutrient removal [8, 9]. As a result of specific operational conditions, such as the selection mechanism favouring granules over flocculant sludge, the AGS technology led to a granular and denser biomass formation [9]; this was favourable for the solid phase separation because no clarifiers are needed [10-12]. Because of the low footprint, lower energy consumption and easy operability than CAS systems, the AGS is getting more and more international attention and imitation [13]. The system comprises one tank that works as a sequencing batch reactor (SBR): fill, react, draw and idle.

Granular properties

The AGS systems' biomass is described as stable aggregates of assorted microbial origin [9]. As can be observed in Figure 1.2, granules have a high density, therefore, can settle very quickly (10 to 90 m/h) without coagulation properties [14, 15]. Their average size diameter ranges from 0.2 to 5 mm, and so, the sludge volumetric index (SVI) at 30 min is more or less equal as at 5 min ($SVI_{30} = SVI_5$) [16, 17]. Their spherical configuration facilitates simultaneous organic matter, P and N removal [12, 18, 19].

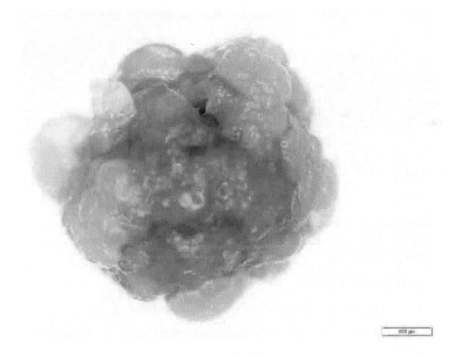

Figure 1.2. Aerobic granule cultured in a laboratory-scale reactor captured by optical microscopy, bar 500 μm.

Source: Md Abdul Hakimvat, IHE-Delft.

3

Biological conversions

In CAS systems, the readily biodegradable organic matter dissolved in the liquid bulk is directly taken up by phosphate accumulating organism (PAOs) during anaerobic conditions and is used for growth under aerobic conditions (anaerobic zone, Figure 1.1). The particulate matter needs more time to become available through hydrolysis. Box 1.1 describes nitrogen-related conversions commonly occurring in biological wastewater treatment processes. Partial or full ammonium oxidation by ammonium-oxidising bacteria (AOB) is the first step of the nitrogen conversion in wastewater treatment plants (WWTP). Denitrification is carried out by nitrite-oxidising bacteria (NOB), in either the pre-denitrification (Pre-D) or the post-denitrification (Post-D) zone of CAS systems (Figure 1.1).

Box 1.1 Nitrogen-related conversions occurring in biological wastewater treatment systems

Nitritation: From NH_4-N to Nitrite (NO_2-N), by AOBs.

Nitrification: From NO_2-N to Nitrate (NO_3-N), by NOBs.

The oxidised forms are typically reduced to dinitrogen gas (N_2) by the ordinary heterotrophic organism (OHO) using organic matter as the electron donor [20]; as follows:

Denitrification: From NO_3-N to N_2;

Denitritation: directly from nitrite (NO_2-N) to N_2; or from

Anammox: NO_2-N and NH_4-N to N_2 carried out by anaerobic ammonium oxidising bacteria.

Source: Unless stated differently, the references are mainly based on Henze, van Loosdrecht [5].

Regarding AGS systems, organic matter, and nutrient conversion occurs in the same granule. The expected nutrient conversions and functionality occurring in the granules are provided in Box 1.2, and is schematically presented in Figure 1.3.

Box 1.2 Chemical conversions occurring in the aerobic granules

In AGS systems, conversions occur in the granules under oxygen-controlled conditions. The microbial diversity in the AGS depends significantly on the granular size and structure [21] and the culture media used [22].

- PAOs are located in the granule's core and can, during the anaerobic feeding period, take up organic matter (feast period) which can be used for growth during aeration. PAOs also positively influence granular density and stability [23].
- Besides the PAOs, also glycogen accumulating organisms (GAO) and NOBs can coexist and are controlled by feast/famine periods [24], temperature [25] and oxygen limitation [26].
  Denitrifying PAOs and denitrifying GAOs can participate in the N and P conversions in the anoxic zones [19, 27, 28].
- Nitrite reduction by denitrifiers occurs in the granule's anoxic zone [27].
- Nitrite oxidation occurs in the outside layer of the granule by NOB, specifically during aeration.
- AOB and OHO grow in the granules' outside layer and can be negatively impacted by high organic loads [29].

5

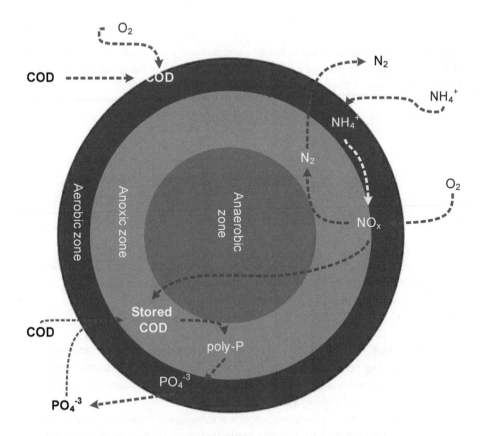

Figure 1.3. Schematic representation of the main process conversions occurring in parallel in the granular sludge structure. Adapted from de Kreuk, Heijnen [30].

1.2 PATHOGENS AND PUBLIC HEALTH – MOTIVATION FOR THIS STUDY

A substantial increase in emerging pathogens has been measured in the last decades [31, 32]. Hazardous water-borne viral and bacterial appearance are related to water contamination and wastewater discharge into the environment. Apart from the requirements for water discharge into the environment, effluent reuse is a practice that has considerable potential for further uptake [33]. However, it may undesirably affect public health. Examples of wastewater as a potential route for spreading pathogenic organisms have been reported. In 2013, the Centers for Disease Control and Prevention (CDC) reported 1.3 million cases of gastroenteritis (diarrheal disease) caused by a norovirus related to effluent discharge into the environment. However, in 1990 the number of cases was 2.6 million. Improvements in water, sanitation and hygiene was mentioned as reason for the reduction over the last 20 years [34]. Still, worldwide, more than 50,000 young people are annually dying by norovirus infections in developing countries [35-37].

Regarding bacteria, local outbreaks of Legionnaires' disease frequently occur. In the USA, more than 10,000 cases are annually reported. *Legionella* can cause severe lung infections that can kill up to 10% of the infected population [34]. Legionella spread through contaminated aerosols that can be conveyed up to 6-10 km, representing a risk for employees and the surrounding communities [38]. Similarly, in the Netherlands, the National Institute for Public Health and the Environment reported an increase of Legionnaires' disease from 2012 to 2017, with about 561 cases. Some community-acquired cases were attributable to surrounded biological WWTPs [39].

Besides viruses and bacteria, other micro-organisms causing infections are the parasites *Giardia intestinalis* and *Cryptosporidium parvum* that can cause giardiasis and cryptosporidiosis, respectively. The CDC attains the contamination by these micro-organisms mainly to poor water treatment and sanitation in developing countries. However, 48% of the globally reported cases (185 outbreaks) were taking place in New Zealand, 154 outbreaks (41%) were seen in the USA, and 9% took place on the European continent; all developing countries [40]. Faecal contamination via treated sewage effluents discharges resulted in local outbreaks [41-43].

Certainly, WWTPs are reservoirs of pathogens. Physical, chemical and biological treatments designed to remove organic matter and nutrients are not sufficient for complete pathogen removal. Apart from the treated effluent, persistent pathogens (bacteria, viruses, and parasites) can leave the WWTP via the primary, secondary (waste) and digested sludge [44-46]. In the case of the promising aerobic granular sludge (AGS) technology, little attention has been given to the pathogen removal process performance. A better understanding of how pathogens are removed may help complementary design measures and reduce the risk of contamination. Due to the large number of emerging pathogens, their dependence on seasonality and local infection, using faecal indicators is recommended.

1.2.1 Faecal indicator of pathogenic organisms used to analyse water quality

The occurrence of pathogens in water has been mostly determined by following surrogates/indicator organisms [47, 48], since they are easier to detect and quantify [33, 49, 50]. Box 1.3 describes the most common organisms linked to faecal water contamination, mostly used in regulations for water reuse and discharge.

Box 1.3 *Definition of the model organisms used as indicator/surrogates or pathogenic organisms in this dissertation.*

Escherichia coli (E. coli)	*Escherichia coli* is a rod-shaped gram-negative bacterium commonly found in the gastrointestinal tract of all warm-blooded animals. Usually, it is a non-pathogenic organism [5]. However, the enterohemorrhagic *E. coli* might provoke gastroenteritis with severe diarrheal disease and urinary tract distress [51]. So far, *E. coli* is considered the most popular faecal coliform indicator for water contamination considering its specificity, environmental persistence, and robustness for analytical detection methods [47, 52].
Thermotolerant coliform (Ttc)	Ttc is also a gram-negative, rod-shaped bacterium that belong to the total coliform group. Its occurrence is strongly associated with faecal contamination. Frequently, 95% of the Ttc belongs to thermotolerant *E. coli* group [53].
Enterococcus spp.	*Enterococcus spp.* is a gram-positive and facultative anaerobic bacterium related to human infections. It is a good indicator of faecal pollution due to its abundance and resistance to environmental conditions. It grows at temperatures from 10 to 45 °C [54].
Bacteriophages	Bacteriophages such as somatic coliphages, F-specific RNA coliphage (i.e. MS2), and *Bacteroides fragilis* are used as indicators for human viruses. They are good indicators because of their resistance to treatments [55] and survival in the environment [56]. Therefore, they are a stable organism for monitoring the efficiency of WWTPs in removing viruses [57, 58].

1.3 RESEARCH GAPS

The aim of this PhD research was to study the pathogen removal and degradation processes in AGS wastewater treatment technology. Due to the large number of pathogenic organisms that can be encountered in WWTPs, their potential association with seasonality, local infections, among other factors, the use of viral and bacterial indicator is preferred and accepted as method.

This study was motivated based on the following arguments:

• The AGS wastewater treatment technology is marketed as a promising wastewater treatment process with many operational and structural (such as low footprint) advantages. As an emergent technology, knowledge of how the AGS contributes to the sanitation goals and public health is essential. Measuring the removal of enteric bacterial and viral faecal indicators as a function of the operational process stages will help to assess the removal capacity of AGS systems and the feasibility of reuse and the effect of discharging treated effluent to the environment.

• AGS is based on a bacterial agglomeration process which contributes to the granular formation and the simultaneous organic matter and nutrients removal. As in other wastewater treatments, in the AGS technology protozoa are present. The substrate type, such as the particulate fraction, used to feed the system might impact the protozoan community. Following granular formation and protozoa's growth under different feeding conditions may help to better understand the pathogen removal mechanisms in the AGS systems.

• How the pathogen removal processes work in AGS systems has not been described in literature yet. Knowledge of the mechanisms involved in removing pathogens and faecal indicators is essential for effluent enhancement. Studies on non-biological mechanisms such as physical removal and selection pressure will enlarge such interest.

• A composite substrate can develop a complex ecosystem. Municipal wastewater contains many micro-organisms and particulates that can influence their abundance. Nowadays, with up-and-coming techniques, eukaryotic groups (protist, fungi, metazoan, archaea and algae) can be extensively identified. For AGS systems, identifying eukaryotic structures benefit the understanding of the mechanisms involved in the pathogen removal degradation process by identifying the potential structures acting as predators.

1.4 RESEARCH AIM

This research aims to enlighten the mechanisms behind the pathogen removal and degradation processes in the AGS treatment systems. The specific objectives are:

• To quantify the removal of bacterial and viral indicators in full-scale aerobic granular sludge processes and conventional activated sludge systems.

• To evaluate the physical and biological mechanisms behind the removal of bacterial and viral surrogates in laboratory-scale systems where the granular fraction is leading.

- To understand the bacterial agglomeration of the AGS and its relationship with the higher-level organisms using an influent containing particulate material at laboratory scale.

- To describe the aerobic granular ecology by the analysis of eukaryotic structures using high-throughput analysis.

1.5 RESEARCH QUESTIONS

- What is the pathogen removal efficiency of the aerobic granular sludge process? Is it comparable to activated sludge systems?

- How are pathogens being removed in aerobic granular sludge systems? Are they mostly predated, getting enmeshed into the small flocculent fraction, or attached to the granular surface? Are there other factors contributing?

- Which changes can be observed when additional particulate material is added to a stable granular sludge system regarding granulation (based on bacteria agglomeration) and its relationship to higher-level organisms as predators?

- From the protist group, who dominate the different flocculent and granular sludge fractions of the aerobic granular full-scale systems? Do they differ from the conventional activated sludge?

1.6 THESIS OUTLINE

This thesis will describe the following chapters based on the previously presented gaps. A schematic representation is also described in Figure 1.4.

Chapter 2: the removal efficiency of viral and bacterial indicators (*E. coli*, F-RNA bacteriophages, *Enterococcus*, Ttc) in two full-scale AGS systems is determined and compared with two parallel, full-scale CAS systems. Differences between AGS and CAS treatments were established during a sampling campaign of six months (autumn and winter). Statistical analysis and correlation with water quality parameters were performed to identify the differences among treatments.

Chapter 3: the removal degradation mechanisms (adsorption, predation and operational conditions) of viral and bacterial pathogen surrogates (*E. coli* and MS2) were investigated in two laboratory-scale AGS reactors. The abundance of the eukaryotic community in each reactor was studied using 18S rRNA sequence analysis. Predation by attached ciliates protozoa was proven using a fluorescent staining technique. The contribution of attachment of the surrogates on the granular surface and the effect of settling in the organism's removal were also evaluated.

Chapter 4: the effect of a high particulate substrate using synthetic faecal sludge was tested in AGS laboratory-scale reactors on granulation, system performance, and ecology. The behaviour of protozoa organisms was analysed. By comparing these reactors with reactors fed on acetate, differences could be determined.

Chapter 5: eukaryotic structures based on rRNA gene (18S) sequencing were investigated in two full-scale AGS systems. The analysis of operational taxonomic units (OTUs) described the abundance of the eukaryotic structures up to species taxa-level in two different granular segregations (small and large granules). Robust statistical analyses were used to identify the differences among two parallel treatments (AGS and CAS systems) fed with the same municipal raw wastewater in two different WWTPs.

Chapter 6 shows the research's outlook and the concluding remarks from this thesis.

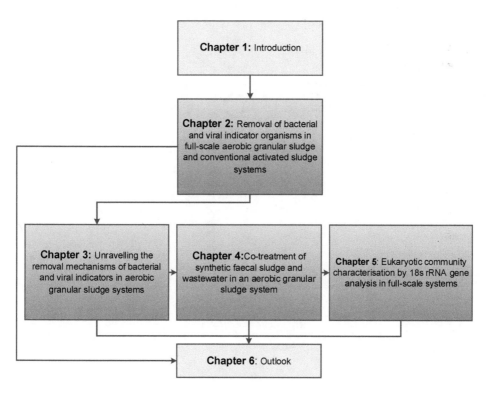

Figure 1.4 Schematic outline of this thesis.

2

FAECAL INDICATORS REMOVALS IN FULL-SCALE AGS AND CAS SYSTEMS

This study contributed to the understanding of the effectiveness of the AGS process in removing faecal indicator organisms compared to the CAS. The work was carried out at two parallel full-scale WWTPS provided with the same influent wastewater - Vroomshoop and Garmerwolde WWTPs. Both systems (CAS and AGS) showed similar FIOs removal efficiency. At the Vroomshoop WWTP, Log_{10} removals for F-specific RNA bacteriophages of 1.4 ± 0.5 and 1.3 ± 0.6 were obtained for the AGS and CAS systems, while at the Garmerwolde WWTP, Log_{10} removals for F-specific RNA bacteriophages of 1.9 ± 0.7 and 2.1 ± 0.7 were found for the AGS and CAS systems. Correspondingly, *E. coli*, *Enterococci*, and TtC Log_{10} removals of 1.7 ± 0.7 and 1.1 ± 0.7 were achieved for the AGS and CAS systems at Vroomshoop WWTP. For Garmerwolde WWTP Log_{10} removals of 2.3 ± 0.8 and 1.9 ± 0.7 for the AGS and CAS systems were found, respectively. Overall, it was not possible to establish a direct correlation between the physicochemical parameters and the removal of FIOs for any of the treatment systems (CAS and AGS). This study opened new research subjects to investigate the mechanisms involved in the FIOs removal.

This chapter is based on: *Barrios-Hernández, M.L., Pronk, M., Garcia, H., Boersma, A., Brdjanovic, D., van Loosdrecht, M.C.M. and Hooijmans, C.M. 2020. Removal of bacterial and viral indicator organisms in full-scale aerobic granular sludge and conventional activated sludge systems. Water Research X 6, 100040*

2.1 INTRODUCTION

Pathogens enter the aquatic environment through municipal wastewater discharges; their occurrence in either treated or raw wastewater may contribute to spreading epidemiological water-borne diseases [40, 59]. Several studies have reported that conventional WWTP do not completely remove pathogens [44, 60-62]. For instance, Lodder and de Roda Husman [35] and van Beek, de Graaf [63] reported the presence of noroviruses and enteric viruses in river basins in the Netherlands originating from treated municipal wastewater discharges.

Microorganisms such as *E. coli* (gram-negative), *Enterococci* (gram-positive) and total coliforms are commonly used as indicators for faecal contamination [64, 65], and as water quality standards for water reuse [33]. Bacteriophages have been used as an indicator for the occurrence of viral pathogens [58, 61, 66]; particularly, the male-specific phages (F-specific RNA bacteriophages) have been used as indicators considering their strong survival in wastewater treatment process and other water environments [67]. Furthermore, F-specific RNA bacteriophages characteristics, such as their isoelectric point, size (22-29 nm), and morphology [35, 68] resemble human viruses such as noroviruses and other enteric viruses [69]. Bacteriophages are also easier to detect than human viruses; inexpensive conventional analytical methods can be used to determine them.

The municipal wastewater treatment worldwide, and particularly in The Netherlands, where this research was carried out, aims at removing organic matter and nutrients [70]. The CAS system is one of the most commonly applied technologies. CAS systems can achieve over 90% removal of C, P, and N. Moreover, CAS systems are effective in removing pathogens; Matthews, Stratton [71] and Amarasiri, Kitajima [66] reported removal efficiencies of faecal indicator organisms (hereafter called FIOs) such as F-specific RNA bacteriophages and total coliforms of 99.5 % and higher. FIOs may be either physically removed, being enmeshed into the flocs during the sedimentation process [69, 72, 73], or biologically predated by other high order organisms such as protozoa [74]. Instead, according to Pronk, de Kreuk [75], full-scale AGS systems may reach C, P, and N removals on average of 87, 86, and 86 %, respectively. However, the performance of full-scale AGS systems on pathogen removal is still unknown.

This chapter compares the removal of FIOs in two full-scale WWTPs in the Netherlands provided with both AGS and CAS systems operated in parallel. The AGS and CAS systems were evaluated on their removal of F-specific bacteriophages, *E. coli*, *Enterococci* and thermotolerant coliforms (TtC). Standard water quality parameters such as chemical organic matter, ammonium (NH_4-N), biological organic matter (BOD_5), orthophosphate (PO_4-P), total suspended solids (TSS), and their relation with the removal of the microbiological organisms, was evaluated as well, in order to see whether water quality parameters can be used to predict FIOs removals.

2.2 MATERIALS AND METHODS

2.2.1 Treatment facilities

This research was performed at two different WWTP in The Netherlands, Vroomshoop and Garmerwolde (Figure 2.1). The WWTPs were initially designed as CAS systems for treating municipal wastewater; later on, AGS systems were incorporated in both WWTPs. Both processes at the two WWTPs (CAS and AGS) work in parallel receiving the same wastewater; moreover, both of them are lacking primary and tertiary treatment (Figure 2.1).

The CAS systems operate in a continuous flow mode, and were designed to remove organic matter (C) and nutrients (N and P). The Vroomshoop WWTP was designed as a biological carrousel CAS system, while the Garmerwolde WWTP as a two stage AB (adsorption/bio-oxidation) CAS configuration with ferric iron dosing in the unit A for P removal. The AGS systems located at both WWTPs consist of a buffer tank (designed to store the influent raw wastewater), followed by a biological tank, containing the granular biomass where all conversions simultaneously occur. The AGS systems are operated as sequencing batch reactors. Figure 2.1 describes the operational conditions for each treatment system at each WWTP.

2.2.2 Sample collection

Grab samples were collected once a month on two consecutive days from December 2017 to May 2018 for each WWTP at three different sampling points: (1) influent wastewater; (2) CAS treated effluent; and (3) AGS treated effluent. The influent wastewater samples, (shown in Figure 2.1 as influent-1) represent the raw municipal wastewater after passing through a grit removal process, and before reaching either the CAS system, or the AGS system's buffer tank. The CAS treated effluent samples (shown in Figure 2.1 as effluent CAS-2) were taken directly from the overflow weir of the CAS settling tank. The AGS treated effluent samples (shown in Figure 2.1 as effluent AGS-3) were collected from the effluent discharge channel after completing one entire batch cycle. The water temperature during the sample campaign ranged between 9 °C and 15 °C.

2.2.3 Sample analysis

The influent and effluent samples were stored in plastic bottles, and were shipped in dark containers refrigerated at 4 °C within 24 hours to two different laboratories as follows. Bacteriophages were determined at the WNL laboratory (Glimmen, the Netherlands). Other microbiological (*E. coli*, TtC, *Enterococci*) and physicochemical parameters (COD, BOD_5, NH_4-N, PO_4-P, and TSS) were determined at the ALS laboratory (Prague, Czech Republic).

15

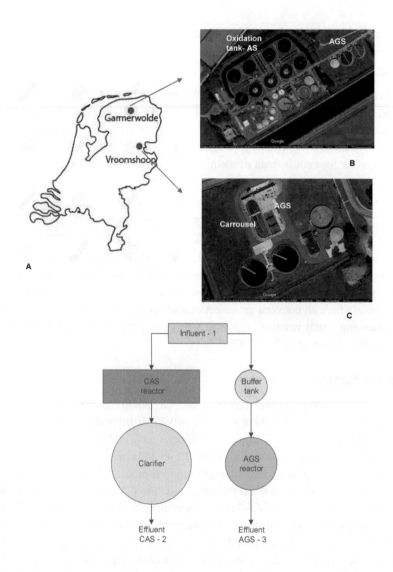

Figure 2.1. Scheme of the combined wastewater treatment plants at Vroomshoop and Garmerwolde WWTPs.

(A) Location of the WWTPs in the Netherlands, Garmerwolde (53°14'51.8"N 6°40'19.6"E) and Vroomshoop (52°26'49.0"N 6°33'52.6"E). Existing facilities, (B) Garmerwolde WWTP containing the Adsorption/Bio-oxidation (A&B) system and the AGS biological tanks. (C) Vroomshoop WWTP, containing a carrousel configuration for the activated sludge treatment and the AGS biological tank.

Water temperature and dissolved oxygen concentrations were collected directly from the WWTPs at the same time the samples were taken. The determinations of the physicochemical parameters were carried out according to the standard methods APHA [76] for the examination of water and wastewater. The determination of the microbiological parameter is explained below.

2.2.4 Bacteriophages enumeration

F-specific RNA bacteriophages were enumerated in duplicate following the double layer method according to ISO 10705-1 [77]. The samples were 10, 100, and 1000-folds diluted in a saline water solution, and then one (1) mL of sample was mixed together with one (1) mL of three hours cultured bacterium host *Salmonella typhimurium* strain WG49. Each sample was mixed in a semi-solid nutrient agar, and poured in a solid nutrient agar plate. Samples were then incubated at 37°C for 18 hours. For enumeration, each plaque in the bacterial mat was counted as one bacteriophage unit. F-specific RNA bacteriophages were determined by plaque-forming units (PFU); the detection limit was reported at 1 PFU/100 mL.

2.2.5 Bacteria enumeration

The analytical determinations of *E. coli* and TtC bacteria were performed following the membrane filtration technique for enumeration according to the standard CSN 75 7835, which is a modified method of the standard method ISO 9308-1 [78] for samples with excessive growth of the accompanying microflora. For enumeration, samples were 10, 100, and 1000-folds diluted in a phosphate-buffered saline (PBS) solution. Samples were plated on *Chromocult®* (Merck Millipore) medium agar for *E. coli* detection and plated in 4-methylumbelliferyl-beta-D-glucuronide *E. coli* broth for TtC detection. Later, they were cultivated 24 hours at 37 ± 2 °C and 44 ± 2 °C for *E. coli* and TtC, respectively.

Enterococci detection and enumeration was performed by the same membrane filtration method according to the standard ISO 7899-2 [79]. The samples were cultivated in the Slanetz and Bartley Medium (Oxoid™) for 24 hours at 44 ± 2 °C. Confirmatory colour reaction tests were also performed to discard false positive by placing the samples in a Bile Esculin Agar (Sigma-Aldrich®, Germany).

All bacteria (*E. coli,* TtC and *Enterococci*) were enumerated by agar plate colony forming units (CFU); the detection limit was reported at 1 CFU/100 mL.

Table 2.1. Operational conditions of the two treatment systems (CAS and AGS) at the Vroomshoop and Garmerwolde WWTPs during the sampling campaign - December 2017 to May 2018

Parameters	Vroomshoop		Garmerwolde	
	CAS	AGS	CAS	AGS
Average Dry flow (ADF) (m³/d)	2,140	1,541	27,645	20,355
Average peak flow (m³/h)	1,125	140	7,148	4,200
Population equivalent COD based	13,100	9,500	210,000	140,000
Organic loading rate (kg COD/m³/d) (aerated)	0.50	0.65	0.53	0.50
Mixed liquor suspended solids range (g/m³)	3 - 6	9 - 14	5 - 8	10 - 14
Aerated basin, O_2 (g/m³)	0.3 - 1.8	1.5 - 2.8	1.0 - 2.0	0.2 - 2.4
Water temperature (°C) range	8.5 - 17.8	8.5 - 17.8	8.6 - 18.0	8.6 - 8.0
Hydraulic retention time (HRT) at ADF(h)	27	11 - 24	24	10 - 12
Sludge retention time (d)	27	> 21	23	> 30

2.2.6 Data analysis

A statistical analysis was performed on the twelve samples collected at each WWTP for each target organism. Data is presented in Box-Whisker plots in which the horizontal line across each box represents the median, the interquartile ranges (50% of the score of the data) and the outliers represent the confident limit of 95% of the determined concentrations. FIOs concentrations were converted to Log_{10}. The removal efficiencies (Log_{10}, %) were calculated considering the concentration of the influent wastewater reaching the WWTPs and the treated effluent discharges for each specific treatment process (AGS and CAS).

Shapiro-Wilk normality test was applied to the converted Log_{10} concentrations of the target FIOs, to the unconverted data of the studied physicochemical parameters and their corresponded removal efficiencies. The Log_{10} FIO removal showed to be normally distributed, therefore the two-paired Student's t-distribution (t-test 95% confidence) analyses were performed to check whether statistically significant differences on the FIOs removal efficiencies exist, on (i) similar processes (that is, CAS at Vroomshoop versus CAS at Garmerwolde, and AGS at Vroomshoop versus AGS at Garmerwolde); and (ii) on different processes at each WWTP (that is CAS versus AGS at both Vroomshoop and Garmerwolde WWTPs). A p-value ≤ 0.05 was used to bind the statistical significance.

Moreover, Pearson's product correlation analyses were performed to determine the possibility for establishing potential trends among the FIOs removal efficiencies; and the Spearman's rank relationship was applied to the comparison of (i) the FIOs and physicochemical concentrations of the influent wastewater, (ii) the FIOs removal efficiency with the removal efficiency of the target standard water quality parameters. The statistical analysis was carried out to determine the presence of any association by calculating the product correlation value from -1 to 1. Moreover, the p-value for each correlation was calculated to determine whether the association was significant (p-value $< = 0.05$) or not. Calculations were statistical computing using R [80].

2.3 RESULTS

2.3.1 FIOs concentrations in raw and treated wastewater

Figure 2.2 and Figure 2.3 show Box-Whisker plots of the FIOs concentrations obtained from each sampling point at the Vroomshoop and Garmerwolde WWTPs, respectively. This corresponds to the microorganism concentration in the raw wastewater (influent to the plant) and the treated effluent from each treatment system (AGS and CAS).

Regarding influent flow wastewater, F-specific RNA bacteriophages were detected in both WWTPs at an arithmetic mean concentration of 10^6 PFU/100 mL; *Enterococci* were detected at mean concentrations of 10^5 CFU/100 mL; and TtC and *E. coli* bacteria both at the same order of magnitude (10^6 CFU/100 mL) in both WWTPs. The average concentrations for each of the evaluated FIOs were of similar magnitude for the two WWTPs. However, no significant correlation was found between the occurrences of the four groups of organisms (p-value > 0.05).

The main variations in the FIOs concentrations observed during the sampling campaign (minimum and maximum values shown in Figure 2.2 and Figure 2.3) can be explained by the hydraulic and seasonal variations when conducting this evaluation. The daily influent flow rate of both WWTPs fluctuates considerably between rain and dry weather flow. At the Vroomshoop WWTP, the minimum influent concentrations for all the target organisms were approximately between one to three orders of magnitude lower than the average concentrations. These lower values were measured in samples that were diluted with rain water because they were taken during rainy weather flow (RWF) events, with a flowrate of around 800 m³/h being more than 63% of the operational design average peak flow (Table 2.1). Moreover, variations from one to four orders of magnitude were observed at the Garmerwolde WWTP. High values were measured in samples taken at a dry weather flow period with a reported flowrate of 2300 m³/h, corresponding to only 15% of the operational design average dry flow at Garmerwolde WWTP.

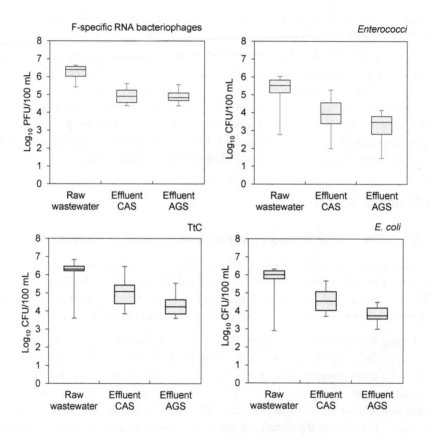

Figure 2.2. FIOs concentrations per sampling point at Vroomshoop WWTP, The Netherlands (December 2017 – May 2018); number of samples n=12.

Although fluctuations in the influent wastewater flow were observed, which had an effect on the concentrations of the target organism, the correlation was not significant. Moreover, also no significant correlations were found between the water temperature (that gradually increased from 9 °C in winter to 15 °C in spring), and the concentrations of the target organisms.

The mean concentrations of the target organisms in the treated effluent of the different processes (AGS and CAS) were not significantly different for the two studied WWTPs. For Vroomshoop the average concentrations of the different FIOs in the AGS treated effluent were slightly lower than the CAS treated effluent. However, the minimum and maximum concentrations showed a high variation. For Garmerwolde, the average concentration of the different FIOs were of the same order of magnitude in the AGS and CAS treated effluent. Minimum and maximum followed similar tendency as at Vroomshoop.

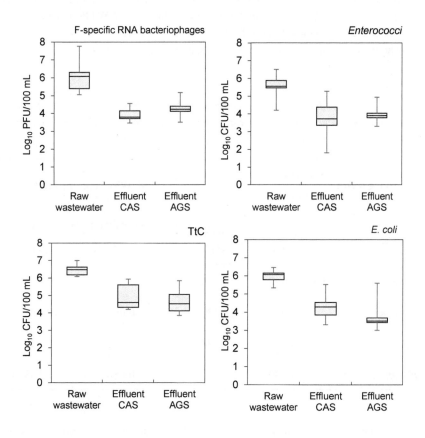

Figure 2.3. FIOs concentrations per sampling point at Garmerwolde WWTP, The Netherlands (December 2017 – May 2018), number of samples n=12.

2.3.2 Log₁₀ removal

The average Log_{10} removals and the standard deviations for the different evaluated FIOs per treatment plant are shown in Figure 2.4. The p-values obtained from the statistical analysis are presented in Table 2.2. Log_{10} removal values for F-specific RNA bacteriophages of 1.34 ± 0.60 (95.5% removal) and 2.13 ± 0.69 (99.3% removal) were reported for the CAS carrousel configuration at Vroomshoop and for the AB-CAS configuration at Garmerwolde, respectively. This difference is statistically significant (p-value = 0.007; < 0.05). No significant statistical differences (p-value > 0.05) could be reported for the rest of the evaluated FIOs between the different CAS systems. *E. coli* average Log_{10} removals of 1.12 ± 0.69 (94.4% removal) and 1.65 ± 0.68 (97.8% removal) were reported at the Vroomshoop and Garmerwolde CAS WWTPs, respectively. Correspondingly, values of 1.60 ± 0.85 (97.3% removal) and 1.88 ± 0.80 (98.7% removal) were reported for TtC at Vroomshoop and Garmerwolde CAS-WWTPs, respectively; and 1.33 ± 0.69 (95.5% removal) and 1.92 ± 0.65 (98.8% removal) for *Enterococci* at Vroomshoop and Garmerwolde CAS WWTPs, respectively.

Table 2.2. p-values from t-test (95% confidence) of the comparison between similar wastewater treatment processes. p-values and R-values from the comparison between different the processes at the each WWTP.

Microorganism	Similar	processes	WWTP comparison	
	comparison			
	CAS vs. CAS	AGS vs. AGS	Vroomshoop CAS-AGS	Garmerwolde CAS-AGS
F-specific RNA bact.	0.007	0.069	0.868	0.405
Enterococci	0.073	0.336	0.069	0.372
TtC	0.623	0.284	0.077	0.585
E. coli	0.239	0.383	0.086	0.543

Regarding the AGS treatment systems, no significant statistical differences (p–value > 0.05) were reported when comparing the Log_{10} removal of the evaluated FIOs at the AGS systems located at the two evaluated WWTPs (Vroomshoop versus Garmerwolde). F-specific RNA bacteriophages Log_{10} removals of 1.38 ± 0.50 (95.8%) and 1.88 ± 0.74 (98.6%), *E. coli* Log_{10} removals of 1.29 ± 0.64 (94.9%) and 1.86 ± 0.94 (98.6%), TtC Log_{10} removals of 2.26 ± 0.78 (99.5%) and 2.05 ± 0.69 (99.0%), and *Enterococci* 1.96 ± 0.47 (98.9%) and 1.67 ± 0.66 (97.9%) were measured for the Vroomshoop and Garmerwolde WWTPs, respectively. Comparing the two different wastewater treatment processes (AGS versus CAS), no significant statistical differences (p-value > 0.05) were observed for the removal of the FIOs by the two treatment technologies at the two evaluated WWTPs.

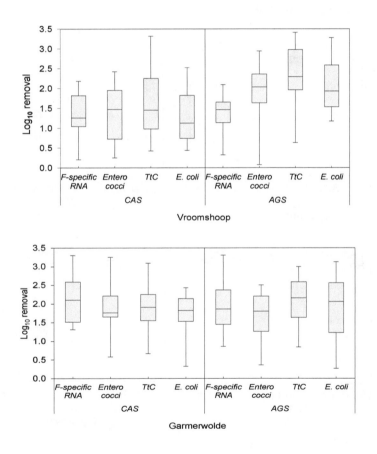

Figure 2.4. Log_{10} removal per treatment plant for AGS (rigth) and CAS (left) systems.

23

2.3.3 Correlation between microbial organisms and water quality related parameters

Except for TSS at Vroomshoop, F-specific RNA bacteriophage concentrations in the influent wastewater at both WWTPs significantly correlated with the concentrations of the measured physicochemical water quality parameters. The rest of the bacteria indicators showed better correlations with the physicochemical parameters at Garmerwolde WWTP compared to Vroomshoop. Regarding the FIOs removal efficiency, the F- specific RNA bacteriophage removal measured at Garmerwolde WWTP significantly correlated with TtC removal (p =0.03) in the CAS system, and with *Enterococci* removal (p = 0.0004) in AGS system, but no significant correlations were found at Vroomshoop WWTP.

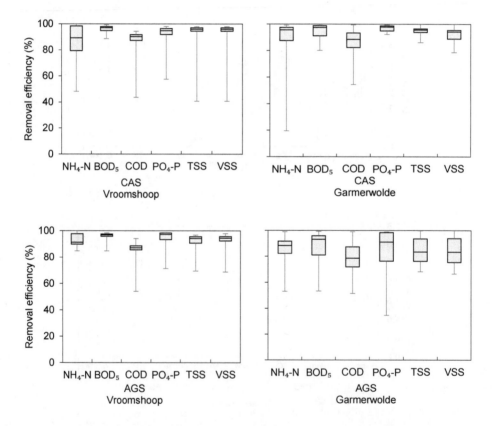

Figure 2.5. Physicochemical water quality parameters removal efficiency (%) for the two different process (CAS and AGS) at each WWTP.

Additional correlation products and p-values can be found in Section 0. Figure 2.5 shows the water quality parameters (NH$_4$-N, BOD$_5$, COD, PO$_4$-P and TSS) removal efficiency (%) for the two different process (CAS and AGS) at each WWTP.

No significant difference between the WWTPs and processes was measured. The above reported FIOs removal was compared with the measured physicochemical water quality parameters per process (CAS or AGS) in each WWTP, in order to evaluate the potential correlation. (Table 2.3 presents the Spearman's rank coefficient (*rho*) and its corresponding p-values. The emphasised values (bold) correspond to statistically significant correlations ($p < 0.05$). Overall, the physicochemical water quality parameters removal efficiencies strongly correlated with each other at both WWTPs (data not shown), but apparently not to all the observed FIOs removal. For the CAS process at Vroomshoop, F-specific RNA bacteriophages showed a significant correlation with most of the studied physicochemical parameters. Moreover, both the *Enterococci* and the *E. coli* removal had a significant correlation with NH$_4$-N, PO$_4$-P and TSS. Conversely, TtC removal showed no substantial association with any of the measured parameters. For the Garmerwolde CAS process, a significant correlation was observed between F-specific RNA bacteriophages and the chemical PO$_4$-P removal and the TSS. *Enterococcus,* TtC and *E. coli* did not significantly correlate with any of the studied parameters

For the AGS system located at Vroomshoop, the F-specific RNA bacteriophages removal showed a significant correlation with the BOD$_5$, PO$_4$-P removals, and it was the only indicator affected by water temperature changes (Table 2.3). Except for *E. coli*, which was positively correlated with the PO$_4$-P, any bacteria indicator significantly correlated with the removals of any of the studied water quality parameters. For the AGS at Garmerwolde, F-specific RNA bacteriophages and *Enterococcus* were positively correlated with the chemical parameters NH$_4$-N, BOD$_5$, COD, and PO$_4$-P, while *E. coli* showed a positive correlation with the physical parameter TSS. TtC were associated with NH$_4$-N, COD, BOD$_5$, and it was the only indicator that showed to be significantly affected by the dissolved oxygen concentration.

Table 2.3. Spearmen rank correlation obtained by correlating the removal of FIOs and the physicochemical parameters per WWTP.

CAS		Vroomshoop				Garmerwolde			
		F- RNA	Enterococci	TtC	*E. coli*	F-RNA	Enterococci	TtC	*E. coli*
NH$_4$-N	rho$_{CA}$s	0.678	0.741	0.497	0.608	0.062	0.440	0.727	0.161
	p-value	0.019	0.008	0.104	0.040	0.851	0.154	0.010	0.619
BOD$_5$	rho$_{CA}$s	0.478	0.384	0.336	0.413	0.559	-0.427	0.210	0.126
	p-value	0.115	0.218	0.286	0.185	0.062	0.169	0.514	0.700
COD	rho$_{CA}$s	0.009	0.245	0.140	0.181	0.552	-0.441	0.175	0.214
	p-value	0.030	0.444	0.670	0.573	0.067	0.154	0.588	0.499
PO$_4$-P	rho$_{CA}$s	0.713	0.657	0.420	0.755	0.664	-0.448	0.056	0.203
	p-value	0.012	0.024	0.177	0.007	0.020	0.147	0.869	0.528
TSS	rho$_{CA}$s	0.734	0.720	0.503	0.671	0.671	-0.224	0.231	0.385
	p-value	0.009	0.011	0.099	0.020	0.020	0.485	0.471	0.218
Temp.	rho$_{CA}$s	0.573	0.259	0.266	0.520	-0.357	0.441	-0.175	-0.046
	p-value	0.055	0.417	0.404	0.080	0.254	0.151	0.586	0.888
DO	rho$_{CA}$s	0.140	-0.007	0.189	0.077	0.213	-0.282	0.450	-0.153
	p-value	0.667	0.991	0.558	0.817	0.505	0.374	0.142	0.636

Continuation Table 2.3. Spearmen rank correlation obtained by correlating the removal of FIOs and the physicochemical parameters per WWTP

AGS		Vroomshoop				Garmerwolde			
		F- RNA	*Enteroc occi*	TtC	*E. coli*	F- RNA	*Enteroc occi*	TtC	*E. coli*
NH$_4$-N	rho$_C$ AS	0.238	-0.329	-0.335	-0421	0.839	0.797	0.622	0.343
	p-value	0.457	0.297	0.287	0.173	0.001	0.003	0.035	0.301
BOD$_5$	rho$_C$ AS	0.531	-0.070	0.153	0.245	0.615	0.650	0.713	0.636
	p-value	0.049	0.834	0.635	0.444	0.037	0.026	0.012	0.030
COD	rho$_C$ AS	0.329	0.392	0.573	0.426	0.643	0.685	0.685	0.503
	p-value	0.297	0.210	0.055	0.169	0.028	0.017	0.017	0.099
PO$_4$-P	rho$_C$ AS	0.664	0.356	0.056	0.636	0.755	0.566	0.716	-0.133
	p-value	0.002	0.256	0.869	0.030	0.006	0.049	0.119	0.683
TSS	rho$_C$ AS	0.301	-0.392	-0.168	0.287	0.455	0.426	0.336	0.755
	p-value	0.343	0.210	0.604	0.366	0.140	0.169	0.287	0.007
Temp.	rho$_C$ AS	0.671	0.063	-0.392	0.427	-0.140	-0.312	0.312	0.154
	p-value	0.020	0.852	0.210	0.169	0.664	0.324	0.324	0.633
DO	rho$_C$ AS	0.007	-0.245	-0.007	0.119	0.385	0.455	0.615	0.098
	p-value	0.991	0.444	0.991	0.716	0.218	0.140	0.037	0.766

2.4 DISCUSSION

The main objective of this chapter was to compare the removal of faecal indicators in AGS systems and CAS systems. To this purpose, detection and enumeration of F-specific RNA bacteriophages, *E. coli*, *Enterococci* and TtC were analysed in raw and the treated wastewater of two full-scale WWTPs in the Netherlands.

The CAS and AGS systems studied were both engineered for domestic wastewater treatment. They manage to efficiently remove representative water quality parameters such as organic matter, TSS and total nitrogen and phosphorous, this in line with the EU

Water Directive (91/271/EEC) on Urban Waste Water Treatment. CAS systems are based on interaction between bacteria and sewage with relatively large land requirements for biological treatment, sludge recycling and separation [5]. The better understanding on the interaction between substrate and bacteria led the development of the AGS system in which denser granular biomass is formed and simultaneously remove organic matter and nutrients in only one biological tank [30]. This improvement has resulted in energy savings (20- 50 %) while compared to CAS systems [81] and other technologies such as membrane bioreactors [82]. Moreover, less area is required compared to CAS systems or other technologies such as trickling filters and constructed wetlands [83]. Those processes normally include primary treatment and in the case of CAS a secondary sedimentation tank after the biological treatment to separate sludge from the liquid bulk. In this study, both studied WWTPs Vroomshoop and Garmerwolde lack primary settling tanks. Therefore, for the comparison of the FIOs removal, it was only needed to consider the different biological treatment and different separation processes, which is either through the settling tanks in the CAS systems or in the same tank for the AGS systems.

2.4.1 FIOs concentrations in raw and treated wastewater

The expected concentrations of F-specific RNA bacteriophages in raw wastewater has been reported to range approximately from 10^4 to 10^6 PFU/100 mL [55, 58, 84, 85]. The concentration of faecal coliforms (*E. coli*, TtC and *Enterococcus*) in raw wastewater ranges from 10^6 to 10^7 CFU/100 mL [5]; Chahal, van den Akker [72] reported for *E. coli* concentrations of 10^4 to 10^9 CFU/100 mL, and Matthews, Stratton [71] reported for *Enterococcus* from 10^5 to 10^6 CFU/100 mL. The arithmetic mean concentrations of those types of organisms in raw wastewater analysed in the current study agrees with previous studies.

2.4.2 Log$_{10}$ removal

The Log$_{10}$ removal of F-specific RNA bacteriophages have been reported to range from 1.5 up to 2.8 in CAS systems [86, 87]. The results obtained in our study are in accordance with those reported values. Regarding the faecal indicator bacteria, the determined Log$_{10}$ removal values for the CAS systems in our research at both Vroomshoop and Garmerwolde WWTPs ranged from 1.1 to 1.9, slightly lower than the values reported in the literature. The microbiological methods are standardized, therefore the differences between the reported data and the present study might be because our CAS systems are compared with CAS systems which have primary and sometimes tertiary treatment. For example, Ottoson, Hansen [44] reported removal values of 3.2 ± 0.8, 3.2 ± 0.1 and 3.5 ± 0.85 for *E. coli*, *Enterococcus* and F-specific bacteriophages in a CAS system with primary sedimentation and sand filters. Little is known about the specific contribution per unit on the removal efficiency. However, not strong contribution of FIOs removal was observed after a pre-treatment stage being an underground septic tank [82] and a primary sedimentation tank [83]. Therefore, the sand filtration unit must have increased the FIOs removal rates in case of Ottoson, Hansen [44].

In this study, the only significant difference in FIOs removal measured, when comparing the CAS process at Garmerwolde WWTP with the CAS process at Vroomshoop WWTP, was for the F-specific RNA bacteriophages (2.1 Log_{10} versus 1.3 Log_{10}). However, when looking at Figure 2, the average values for all FIOs removal were higher for Garmerwolde CAS process compared to Vroomshoop. This can be explained by reviewing the pathogen removal mechanisms in wastewater treatment systems. It may be expected that due to the FIOs isoelectric point, they get attached to the flocs and then end up the sludge line - with a high probability to remain alive in the sludge cake [73]. Due to the ferric iron dosing in the unit A, which contributes to the floc formation, the two stage AB-CAS process in Garmerwolde WWTP might enable extra FIOs removal through flocculation. Vroomshoop is operated as a carrousel and does not have ferric iron dosing. Moreover, the CAS system at Vroomshoop WWTP receives the excess/waste sludge from the AGS system [81], which may result in an additional accumulation of FIOs in the biological tank. In Garmerwolde WWTP, the excess sludge is directly discharged to the sludge digestion line.

Another removal mechanism in CAS systems, such as biological predation in the aeration tank by protozoa organisms, have been broadly studied [88, 89]. Predation might equally affect bacteriophages and faecal bacteria, but an extra biological removal mechanism might occur for bacteria such as cell lysis using *E. coli* or *Enterococcus* as preferred organism for human virus and bacteriophages replication [90]. Reproduction of the host bacteria is a prerequisite to get infected by bacteriophages [68].Since the experimental period of this study was carried out during winter and spring with a low range of temperature, bacteria lysis by bacteriophages replication likely not contributed in the removal. In line with Wen, Tutuka [46] findings, the overall low bacterial and viral indicators removal efficiencies found in this study could have been influenced by the low temperatures in which the campaign was carried out (< 15 °C). The lower the temperature, lesser number of bacteria is expected to be adsorbed into a solid phase. Those effects can be due to changes of the water/microbial surface viscosity, reduction of some chemical/physical adsorption properties and changes physiology of the organisms [91, 92].

With respect to the two AGS systems (Vroomshoop and Garmerwolde WWTPs), both systems showed a similar FIOs removal efficiency. Results indicate that the differences in the flow treated and the treated organic loading rate had little impact on the FIOs removal. When comparing AGS with CAS systems, the FIOs removal in the two AGS systems showed to be statistically similar as in the CAS systems for the two WWTPs. Our results are in accordance with a previous study conducted by Thwaites, Short [85] who compared the removal of related FIOs in pilot scale AGS and CAS systems. The design and operational conditions of the full-scale AGS systems such as absence of primary treatment, solid/liquid separation occurring in the same biological reactor, a short sedimentation period and shorter HRT than in the CAS systems may have an effect on the FIOs removal mechanisms. However, although different mechanisms might be taking place in the CAS and AGS systems, the overall amount of FIOs removal was found to be similar. The operational conditions of an AGS system determine the formation of a spherical shaped granule which surface can function as adsorption area for FIOs. Little is known about whether FIOs can be adsorbed or attached to the surface of the granular biomass. Thwaites, Short [85], applied a method to compare the detachment of FIOs from particulate materials in CAS and mature AGS. Contrary to the F- specific RNA bacteriophages, their results showed larger separation of sulphite-reducing clostridia, *E. coli* and total coliforms from the AGS than CAS. The applied technique was not selective enough to distinguish which mechanism (absorption or single adhesion to the granule surface) was involved, thus additional studies were recommended to understand the contribution of the granular composition in the FIOs removals.

Moreover, attachment of ciliate protozoa on the surface area of the granules can also take place, facilitating biological predation of the FIOs by the protozoa [85, 93, 94]. The high biomass density in the AGS systems can also be a limitation for bacteriophages propagation since the granules may provide shelter to host bacteria [95]. Therefore, further studies are needed to better understand the AGS process and how it relates to the pathogen removal mechanisms.

2.4.3 Correlation between microbial organisms and water quality related parameters

Correlations between bacteriophages, bacteria indicators and physicochemical parameter removal and wastewater influent concentrations reported in this study are in accordance with previous literature [55, 58, 82]. In agreement with De Luca, Sacchetti [55], the physicochemical parameters measured in this study better correlated with each other than with the target microorganism removals. Ottoson, Hansen [44], [45] reported removal trends between FIOs COD and total organic carbon in a CAS system. For the AGS process at Garmerwolde, correlations between the removal of the FIOs and COD, BOD_5 and NH_4-N were found. *E. coli* removal strongly correlated with the TSS removal in the AGS system at Garmerwolde WWTPs. According to van Dijk, Pronk [96], the rapid liquid bulk separation from the biomass carried out in the biological tank of the AGS system causes a disturbance of the biomass during feeding which lead to wash-out of particles measured as TSS in the effluent, which might also include FIOs. The rest of the measured physicochemical water quality related parameters were randomly correlated with the FIOs removal for both treatment systems, and could not be used to predict FIOs removal in this study.

2.5 CONCLUSIONS

The faecal indicator (F-specific RNA bacteriophages, *E. coli*, *Enterococci*, and TtC) concentrations in the raw influent wastewater of two WWTPs in the Netherlands, Vroomshoop and Garmerwolde were found to be comparable with those reported in literature. On average FIOs (statistically significant for F-specific RNA bacteriophages only) were removed less efficiently in a simpler CAS configuration (carrousel at the Vroomshoop WWTP) compared to a more complex one (AB-CAS system at the Garmerwolde WWTP), which might be related to the sludge separation and discharge. AGS systems remove F-specific RNA bacteriophages, *E. coli*, *Enterococci*, and TtC as efficient as CAS systems when treating the same raw influent wastewater. Measured water quality parameters (NH_4-N, BOD_5, COD, PO_4-P and TSS) could not accurately predict FIOs removal at any of the evaluated treatment systems.

2.6 ANNEX 1

2.6.1 Statistical analyses products

Table 2.4. Spearmen rank correlation obtained by comparing the FIOs concentration at the influent and the physicochemical parameters concentrations per WWTP. Values stand for rho and p-value.

	Vroomshoop				Garmerwolde			
	F-specific RNA	*Entero cocci*	TtC	*E. coli*	F-specific RNA	*Entero cocci*	TtC	*E. coli*
Enterococci	-0.084				0.462			
	0.800				0.134			
TtC	0.399	0.658			0.559	0.483		
	0.199	**0.019**			0.063	0.115		
E. coli	0.410	0.326	0.777		0.582	0.368	0.067	
	0.186	0.302	**0.003**		**0.047**	0.239	0.837	
NH4-N	0.776	0.119	0.398	0.427	0.860	0.601	0.611	0.674
	0.005	0.716	0.240	0.166	**0.001**	**0.043**	**0.020**	**0.016**
BOD$_5$	0.846	0.133	0.350	0.207	0.678	0.608	0.720	0.611
	0.001	0.683	0.264	0.519	**0.018**	**0.040**	**0.011**	**0.035**
COD	0.783	0.105	0.613	0.637	0.762	0.608	0.734	0.533
	0.004	0.750	0.034	0.026	**0.006**	**0.040**	**0.009**	0.074
PO$_4$-P	0.839	0.105	0.434	0.543	0.825	0.545	0.573	0.747
	0.001	0.750	0.158	0.067	**0.002**	0.071	0.055	**0.005**
TSS	0.469	0.161	0.287	0.332	0.622	0.545	0.503	0.646
	0.128	0.619	0.365	0.291	**0.035**	0.071	0.099	**0.023**

Table 2.5. Pearson products and p-values obtained for correlating the bacteriophages removal with bacteria indicator at Vroomshoop WWTP. Values stand for rho and p-value.

	Vroomshoop WWTP					
	CAS			**AGS**		
	F-specific RNA	*Entero cocci*	**TtC**	**F-specific RNA**	*Entero cocci*	**TtC**
Enterococci	0.45			0.43		
	0.14			0.16		
TtC	0.12	0.82		0.05	0.61	
	0.68	**<<0.05**		0.89	**0.04**	
E. coli	0.38	0.82	0.70	0.46	0.69	0.54
	0.78	**<<0.05**	**0.01**	0.14	**0.01**	0.07
	Garmerwolde WWTP					
	CAS			**AGS**		
	F-specific RNA	*Entero cocci*	**TtC**	**F-specific RNA**	*Entero cocci*	**TtC**
Enterococci	0.02			0.85		
	0.95			**<<0.05**		
TtC	0.63	0.29		0.38	0.60	
	0.03	0.37		0.22	**0.04**	
E. coli	0.45	0.66	0.53	0.24	0.35	0.50
	0.14	**0.02**	0.85	0.46	0.27	0.10

3

UNRAVELLING THE REMOVAL MECHANISMS OF FAECAL INDICATORS IN AGS SYSTEMS

This chapter focussed on determining the relation between reactor operational conditions (plug flow feeding, turbulent aeration and settling), and physical and biological mechanisms on removing two faecal surrogates, $E.\ coli$ and MS2 bacteriophages. Two AGS laboratory-scale systems were separately fed with influent spiked with 10^6 CFU/100 mL of $E.\ coli$ and 10^8 PFU/100 mL of MS2 bacteriophages and followed during the different operational phases. The reactors contained only granular sludge and no flocculent sludge. Both systems showed reductions in the liquid phase of 0.3 Log_{10} during anaerobic feeding caused by a dilution factor and attachment of the organisms on the granules. Higher removal efficiencies were achieved during aeration, approximately 1 Log_{10} for $E.\ coli$ and 0.6 Log_{10} for the MS2 bacteriophages caused mainly by predation. The 18S sequencing analysis revealed high operational taxonomic units of free-living protozoa genera $Rhogostoma$, and $Telotrochidium$ concerning the whole eukaryotic community. Attached ciliates propagated after the addition of the $E.\ coli$, an active contribution of the genera $Epistylis$, $Vorticella$, and $Pseudovorticella$ was found when the reactor reached stability. In contrast, no significant growth of predators occurred when spiking the system with MS2 bacteriophages, indicating a low contribution of protozoa on the phage removal.

This chapter is based on: *Barrios-Hernández, M.L., Bettinelli-Travián, C. Mora-Cabrera K., Vanegas-Camero M.C., Garcia, H.A., van de Vossenberg, Prats D., Brdjanovic, D. van Loosdrecht, M.C.M., and Hooijmans, C.M. Unravelling the E. coli and MS2 bacteriophage removal mechanisms in aerobic granular sludge systems. Water Research, 116992.*

3.1 INTRODUCTION

In general, AGS wastewater treatment systems report high removal efficiencies of carbon, nitrogen and phosphorus [75, 81, 97]. Besides a good treatment performance concerning the water quality parameters, two recent studies showed the capability of the AGS technology in removing bacterial and viral indicator organisms from sewage. Barrios-Hernández, Pronk [98] and Thwaites, Short [85] compared removal efficiencies in AGS and CAS full- and pilot-scale WWTP. The AGS full-scale systems can just as effectively remove indicator organisms as the CAS process. For example, the Log_{10} removal for both systems ranged between 1.7 and 2.6 for bacteria as *E. coli*, and between 1.4 and 2.4 for F-specific RNA bacteriophages. Both studies mentioned above, emphasised that the presence of the variety of protozoa commonly present in wastewater treatments could be influencing the removal of the indicator organisms.

For a good understanding of pathogen removal by AGS systems, more mechanistic studies are needed. A large number of studies can be found for other wastewater treatment systems looking at biological (cell lysis and predation) and physical (adsorption and precipitation) removal mechanisms of pathogenic bacteria. A study by van der Drift, van Seggelen [99] postulated that the faecal surrogate *E. coli* was either biologically predated by protozoa, or ended up enmeshed into the sludge flocs. Hereafter, other researchers confirmed the importance of protozoa as grazers in CAS systems [74, 89, 100]; and their role as primary predator during aeration [101]. More recently, the removal of viruses in CAS systems has been studied, using bacteriophages as a surrogate for viruses, showing that their elimination from sewage can be challenging due to their persistence and abundance [60, 66]. Bacteriophages tend to either attach or detach from surfaces depending on the surrounding water conditions [102]. They can also be predated by heterotrophic flagellates [103, 104]. According to Stevik, Aa [92] and Dias, Passos [105], their retention and depletion in wastewater may be affected by system configuration, hydraulic retention time, water quality (temperature, pH and organic matter), and water flow velocity, among other factors.

The main goal of this study was to clarify the removal mechanism in AGS systems, especially in the granular fraction. The association with of the operational conditions of an AGS laboratory-scale reactor with the removal of a faecal bacterial surrogate *E. coli* and a viral surrogate MS2 bacteriophage was evaluated. Attachment of the faecal organisms onto the granules, protozoa predation and the contribution of the settling in the bacterial and viral surrogate removal process was part of the study.

3.2 MATERIALS AND METHODS

3.2.1 Research design

Two laboratory-scale reactors were operated long-term as sequencing batch reactors (SBR). Both systems developed a steady-state situation with mature granules when fed with only synthetic wastewater. When the steady-state was reached, the influent was spiked with known concentrations of two typical surrogates for bacterial and viral water quality, *E. coli* bacteria and MS2 bacteriophages. Reactors were monitored weekly for physicochemical and microbiological water quality parameters, before and after each cycle operational phase (anaerobic plug feeding, aeration phase and settling). Changes in the protozoa community were observed using microscopy observation, and changes in the eukaryotic community were studied using 18S rRNA sequence analysis. Next to the long-term investigation, additional batch experiments were executed to better understand predation (using a fluorescent staining technique) and attachment of the surrogates on the granular surface. The contribution of the settling phase to the removal of the studied surrogates was also evaluated.

3.2.2 Laboratory-scale SBR

Two laboratory-scale SBRs (hereafter called AGS_*E. coli* and AGS_MS2) were operated for 154 and 125 days, respectively. The operational cycles follow the sequence of an anaerobic phase, aeration (reaction) phase, and settling and effluent withdrawal (Figure 3.1). AGS_*E. coli* was controlled with a Braun DCU4 controller, coupled with both mass-flow and a multi-fermenter control system (MFCS), using acquisition software (Santorious Stedim Biotech S.S., Germany). AGS_MS2 was controlled with an Applikon ADI controller model 1030, connected to a computer with the software BioXpert 2 (Applikon, the Netherlands). Both systems were inoculated with crushed sludge from an AGS full-scale WWTP (Garmerwolde, the Netherlands) with an initial total suspended solids (TSS) concentration of 8.3 ± 2.5 g/L.

3.2.3 Synthetic wastewater

The systems were fed with the synthetic wastewater composed of acetate (2.9 kg/m^3/day), ammonium-nitrogen (0.48 kg NH$_4$-N/m^3/day), phosphorus source (0.08 kg PO$_3$-P/m3/day) and trace metals prepared according to the Vishniac and Santer [106]' solution. The bacterial and viral surrogates were added once the granular stability in the reactors was established, i.e., after day 47 in the AGS_*E. coli* reactor and after day 69 of operation in the AGS_MS2 reactor. The concentration in the influent for the AGS_*E. coli* reactor was between 1×10^4 and 1×10^7 CFU/100 mL. For the AGS_MS2 reactor concentrations were between 1×10^5 and 1×10^8 PFU/100 mL.

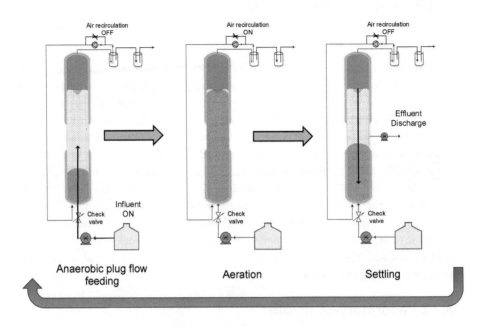

Figure 3.1 Cycle operational conditions of the AGS reactors

3.2.4 Bacterial and viral surrogates

Due to their importance for water quality regulations, two faecal surrogates were propagated as explained in Scoullos, Adhikari [107].

E. coli strain, culture and enumeration

The *E. coli* ATCC reference strain 25922 was taken as bacteria surrogate for faecal contamination. *E. coli* was initially inoculated in a sterilised Nutrient Broth medium (Merck KGaA, Germany) and incubated on a shaking platform (150 rpm) at 37 ± 1 °C for 24 hours. A stock concentration of about $1x10^9$ CFU/100 mL was obtained and enumerated by spreading the medium on *Chromocult* (Sigma-Aldrich, Germany) coliform agar plates. Later, the suspension was spiked to the influent of the AGS_*E. coli* reactor by diluting the stock culture into a vessel of 10 L to an end concentration in the reactor of $1x10^7$ CFU/100 mL. For enumeration, viable counts were conducted in triplicate as described in Section 2.2.5 for *E. coli* bacteria. Aliquots of 0.1 mL of either pure or diluted sample were spread on the coliform agar plates and inoculated overnight at 37 ± 1 °C. Undiluted samples with expected concentrations lower than 30 CFU/100 mL were analysed in duplicate using membrane filtration. That is, 100 mL of the undiluted sample was passed through a cellulose nitrate membrane filter (0.45 μm). The filter was placed on *Chromocult* coliform agar plates and incubated at 37 ± 1 °C for 24 hours.

Bacteriophage strain, culture and enumeration

The *E. coli* bacteriophage MS2 reference strain ATCC 15597-B1 was used as the viral surrogate. The phage was propagated in Tryptone Yeast Glucose Broth (TYGB) using the *E. coli* strain C-3000 (ATCC 15597) as a host bacterium while shaking at 150 rpm. The incubation temperature was 37 ± 1 °C for 24 hours to reach a stock concentration of 1×10^{12} PFU/100 mL. Working solutions were prepared in saline water buffer before being applied to the AGS_MS2 reactor by diluting the stock culture into a vessel of 10 L to a concentration 1×10^8 PFU/100 mL. The MS2 bacteriophage enumeration was determined as explained in Section 2.2.4 for F- RNA bacteriophages, but using the *E. coli* strain C-3000 as a host.

3.2.5 Sample collection and processing

Physicochemical water quality parameters

For the performance of the reactor, 10 mL samples were taken from the liquid bulk before the aeration phase (62 min) and from the effluent. The samples were filtered through a 0.45 μm filter (Millex-HV, Germany) and subjected to the following analysis: COD, orthophosphate (PO_4-P) and nitrogen-related parameters such as NH_4-N, nitrite (NO_2-N), and nitrate (NO_3-N). For the COD measurement, the Closed Reflux-Colorimetric Standard Method [76] was used. The rest of the measurements were performed using LCK (Hach, Germany) cuvette tests. To control the optimal biomass growth, the TSS and volatile suspended solids (VSS) were determined according to the Standard Methods [76] for sludge samples and treated effluent samples.

3.2.6 Microbiological sampling process

For microbiological enumeration, samples were taken weekly from both reactors (AGS_*E. coli* and AGS_MS2) at the following sampling points: influent (10 mL), mixed liquor at the end of the anaerobic phase (25 mL), mixed liquor at the end of the aerobic phase (25 mL) and effluent (10 mL). From the mixed liquor samples, the sludge was separated from the liquid (hereafter referred to supernatant) by letting the sludge settle for 5 min. The supernatant (10 mL) was extracted with a syringe and placed in a separated vessel. Approximately 1 mL of the settled sludge fraction was crushed and homogenised using a glass/Teflon potter Elvehjem tube. All samples were enumerated in duplicate as explained in Section 3.2.4. Results from the supernatant and sludge fractions after anaerobic and aerobic phases were subjected to a statistical analysis (Wilcoxon signed-rank test) after normalization to determine whether paired mean concentrations were significantly ($p <$

0.05) different from each other or not. The number of samples (n) for the analysis was between 8 and 13.

3.2.7 Optical microscope observation of protozoa

Additional samples of granules (5 mL) were taken during aeration to be inspected for protozoa presence. Samples of 25 µL were observed under optical microscopes Olympus CH30 (10x, 20x, and 40x) and Olympus BX51 (10x, 20x and40x). The stalked ciliated protozoa activity (occurrence and mobility) was studied based on a qualitative and quantitative scale observation of the individuals, as described in Amaral, Leal [108]. For the 40x magnification, an area of approximately 37 mm^2 was measured, the highest value of 100% was assigned to the ones that showed high activity and more than six individuals/mm^2 in all the observations. A value of 5% was assigned to samples that at least showed one individual/mm^2 in any of the measured samples. Samples were checked in triplicates.

3.2.8 DNA extraction and 18S rRNA gene sequencing

Genomic DNA was extracted from approximately 0.25 g of crushed sludge collected on days 104 and 160 from the AGS_*E. coli* reactor, day 90 from the AGS_MS2 reactor, and the seed sludge using QIAamp PowerFecal PRO DNA kit (QIAGEN). The DNA concentration was determined using an Invitrogen Qubit Fluorometer (Thermo Fisher Scientific, USA). The V4 region of 18S rRNA genes was amplified using the following eukaryote-specific primers pair 528F 5'-GCGGTAATTCCAGCTCCAA-3' and 706R 5'-AATCCRAGAATTTCACCTCT-3'. PCR reactions were carried out with Phusion High-Fidelity PCR Master Mix (New England Biolabs). 1x loading buffer (contained SYBR green) was mixed with the PCR products and run on 2% agarose gel electrophoresis. Products between 400bp-450bp were purified using the Qiagen Gel Extraction Kit (Qiagen, Germany). The libraries were generated with NEBNext UltraTM DNA Library Prep Kit for Illumina (Illumina NovaSeq 2500, USA) and quantified via Qubit and Q-PCR. Paired-end reads were merged using FLASH (V1.2.7). Chimeras were removed using Qiime (Version 1.7.0), and sequences analysis were performed by Uparse software (Uparse v7.0.1001). Operational Taxonomic Units (OTUs) were obtained by clustering with ≥ 97% similarity. The analysis was performed using Silva database for species annotation. The raw sequence data were uploaded to the National Center for Biotechnology (NCBI) under accession numbers: SAMN16526359, SAMN16526360, SAMN16526361, and SAMN16526362.

3.2.9 *E. coli* fluorescence microscopy observations

To identify and record protozoa predation, a fluorescence staining detection method for *E. coli* was used. Granules from an additional AGS laboratory-scale reactor as well as granules from a full-scale WWTP were checked on the abundance of ciliates attached to the granular surface. The *E. coli* ATCC 25922 was labelled using a dsGreen gel staining solution 10,000x Lumiprobe (Hannover, Germany). It was analysed with the fluorescein isothiocyanate (FITC) filter set in the microscope. *E. coli* was 10-fold diluted as follows, 2 µL of the 10,000x dilution of dsGreen was added to a 1.998 µL of *E. coli* ATCC 25922 to obtain the final working solution of 1×10^4 CFU/ µL. Mini batch reactors were prepared in 2 mL Eppendorf tubes in which 1 mL of granules were spiked with 1 mL of the solution with the previously labelled *E. coli*. The solution was, quickly mixed three times in a pulsing vortex mixer (VWR, Germany), then incubated in the dark for 15 minutes. The treated granules were washed three times in 400 µL of 1x phosphate-buffered saline (PBS) and centrifuged at 4,000 rpm for 5 min (Eppendorf MiniSpin, Germany); then resuspended in 1x PBS to get a final volume of 2 mL. Aliquots of 3 µL were placed on glass slides and analysed under an Olympus BX51 fluorescent microscope coupled with an XM10 camera, an X-cite fluorescence lamp (Lumen Dynamics, Series 120Q) and a FITC filter. Approximately between 8 and 12 sets of pictures of different visual parts of the granules were taken. For each set of pictures, both phase contrast and fluorescence images were taken at magnifications from 10 to 100x. Overlay pictures were analysed using Fiji image analysis software (https://fiji.sc/).

3.2.10   Attachment of *E. coli* and MS2 bacteriophages

To determine whether *E. coli* and MS2 bacteriophages attached to the granules, batch tests were performed at the same temperature as for the long-term study (20 °C). Round-shaped granules (from 0.2 to 3.8 mm) from an additional control AGS laboratory-scale reactor were tested. The AGS reactor was fed only with synthetic wastewater. Therefore, there were no *E. coli* bacteria, MS2 bacteriophages, nor potential predators microscopically detectable such as free-swimming and attached ciliated protozoa. The experimental tests were carried out based on Hendricks, Post [91] with the following modifications. Three beakers were prepared with 50 g of the fresh granules and filled up to 200 mL with a synthetic wastewater solution. The beakers were mixed continuously with a magnet stirrer (250 rpm); then spiked with a known concentration of the target microorganisms (10^5 and 10^7 CFU/100 mL of *E. coli* bacteria, and 10^6 and 10^9 PFU/100 mL of MS2 bacteriophage). The initial concentration in the attachment test (C_0) was measured by taking 1 mL from the suspension (liquid bulk). The experiments aimed to analyse the behaviour of the surrogates during the 60 minutes of anaerobic feeding, therefore, 1 mL was consecutively taken from the suspension after letting the sludge settle at times (C_t) 5, 15, 30, 45 and 60 min. Since target organisms keep suspended in the liquid phase, the difference between C_0 and C_t was assumed to be caused by attachment to the granular media. Experiments were performed in duplicates, as well as the spreading and enumeration of the microorganism.

Kinetics were calculated using the pseudo-second order equation (Eq. 1) explained by Simonin [109], in which t corresponds to the exposure time in minutes between the target organism and the granules, q is the *E. coli* bacteria (CFU/g) or the MS2 bacteriophages (PFU/g) concentration attached per gram of granule, q_e is the maximum attachment capacity of the organisms (CFU/g or PFU/g), and k is the fitted constant.

$$\frac{t}{q} = \left(\frac{1}{q_e}\right)t + \left(\frac{1}{kq_e^2}\right)$$

Equation 1

3.2.11 Contribution of the settling in the AGS reactor

The effects of the settling on the removal of the target microorganisms were independently evaluated in an additional AGS reactor, operated like the long term studied reactors but without being fed with any surrogate. The granules, cultivated only with synthetic wastewater, were spiked with either *E. coli* bacteria or MS2 bacteriophages and thoroughly mixed by aeration for 5 min to reach an equilibrium between supernatant and granules. The test was performed twice with different concentrations per indicator (around 10^5 and 10^7 CFU/100 mL for *E. coli* and around 10^3 and 10^9 PFU/100 mL for MS2) to determine in how far the outcome of the test was affected by the concentration. At the end of the 5 min mixing, a sample was taken to determine the initial indicator concentration in the reactor. After turning off the aeration, a settling time of 5 min was allowed. After settling, the following samples of 5 mL were taken: treated supernatant at three different heights of the reactor column from the effluent discharge point (20, 40 and 60 cm), and a final mixed effluent sample. To separate the liquid fraction from the solids, the mixed sample taken before settling and the effluent were treated by let them settle for 5 min in a measuring cylinder. After that, microbial spreading and enumeration were carried out as described in Section 3.2.4.

3.3 RESULTS

3.3.1 AGS reactors performance

The performance of two granular sludge reactors (AGS_$E.$ $coli$ and AGS_MS2) for the concentration of COD, PO_4-P, NH_4-N, NO_2-N, and NO_3-N are given in Figure 3.2. The reactors contained only granular sludge while flocculent sludge was absent due to feeding with soluble substrate only.

The measurements showed good performance in terms of COD removal, see Figure 3.2, a and b. Concentrations were reduced at the end of the anaerobic phase from 402 ± 50 mg COD/L to averages of 40 ± 22 and 65 ± 17 mg COD/L after 47 and 57 days of operation in the AGS_$E.$ $coli$ and AGS_MS2 reactor, respectively. The effluent showed final average concentrations of 36 ± 21 and 34 ± 14 mg COD/ L, correspondingly. This effluent COD was mainly related to the non-biodegradable EDTA present in the influent. The systems also showed P-release with average values of 59 ± 16 mg PO_4-P/L for AGS_$E.$ $coli$ and 61 ± 22 mg PO_4-P/L for AGS_MS2.

The phosphate removal was always good, with concentrations lower than 1 mg PO_4-P/L in the treated effluent for both reactors. Regarding nitrogen, the average ammonia-nitrogen concentration in the influent was 58 ± 13 mg NH_4-N/L, which was partially converted to NO_2-N and NO_3-N during the aeration phase. Effluent values were on average 20 ± 19 mg NH_4-N /L, 2 ± 2 mg NO_2-N /L, and 4 ± 5 mg NO_3-N /L for the AGS_$E.$ $coli$ reactor (Figure 3.2, c and e).

For the AGS_MS2 reactor average concentrations of 8 ± 9 mg NH_4-N /L, 2 ± 2 mg NO_2-N /L, and 0.6 ± 0.5 mg NO_3-N /L were measured (Figure 3.2, d and f). It seems that copper from the feeding valve negatively affected the ammonia-oxidising bacteria community in the AGS_$E.$ $coli$ reactor. The valve was in use from day 90 to day 148. The dissolved oxygen was increased from 1.8 to 3.8 mg/L to stimulate the nitrification process. However, after replacing the valve, this was not necessary anymore. Oxygen was set at 1.8 mg/L; after which the system stabilised again. Since the behaviour of the N-conversion process was assumed not be influencing the removal of $E.$ $coli$ and MS2 bacteriophage the nitrification was not optimised.

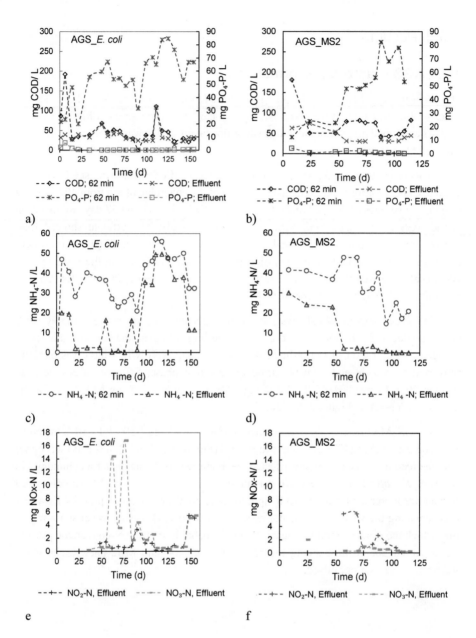

Figure 3.2 Performance of the laboratory-scale AGS reactors fed with E. coli (AGS_E. coli) or MS2 bacteriophages (AGS_MS2) for COD, phosphate and nitrogen removal.

3.3.2 Fate of the faecal surrogates in the long-term AGS laboratory-scale SBRs

The measured median E. *coli* concentration in the influent was 1.0×10^6 CFU/100mL, ranging from 4.5×10^4 to 2.0×10^7 CFU/100 mL in the AGS_E. *coli* reactor. The MS2 bacteriophage influent concentrations for the AGS_MS2 reactor ranged between 4.0×10^5 and 7.5×10^8 PFU/100 mL, with a median of 1.3×10^8 PFU/ 100 mL.

The effluent concentrations were between 1.0×10^2 and 7.1×10^5 CFU/100 mL for E. *coli* (median of 9.0×10^4 CFU/100 mL); and the MS2 bacteriophage concentrations were between 3.0×10^5 and 1.9×10^8 PFU/100 mL (median of 5.5×10^7 PFU/100 mL). The overall median of the E. *coli* and MS2 bacteriophage removal in the systems, comparing influent and effluent, was 2.2 and 0.3 Log10 (Figure 3.3, a), respectively.

E. *coli* and the MS2 bacteriophages removal profiles were fitted to a Chick–Watson model (Figure 3.3, b) using the average concentrations measured in the influent (C_0) and the liquid bulk fractions (C); at the end of both the anaerobic phase and the aerobic phase and effluent. After the anaerobic phase (at time 62 min) a reduction of 0.3 Log10 was measured for AGS_E. *coli* and AGS_MS2 reactors. During aeration, the E. *coli* concentration decreased by 0.9 Log10, whereas the MS2 bacteriophages decreased by 0.6 Log10. After settling time, a further reduction of 0.5 Log10 was measured for E. *coli*. MS2 bacteriophage remained constant.

a   b

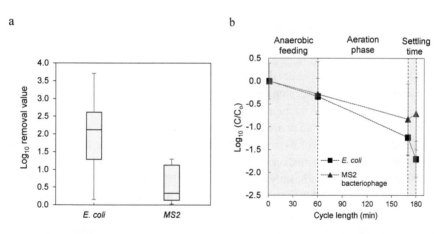

Figure 3.3 Overall average removal of the target surrogates

E. coli bacteria (n =17) and MS2 bacteriophages (n = 9) in laboratory AGS reactors ()
and average depletion curves per operational cycle (b).

3.3.3 Surrogates concentrations in the sludge and liquid fractions per operational phase

Figure 3.4 shows the Log_{10} concentrations of the target microorganisms (*E. coli* bacteria and MS2 bacteriophage) measured after the fractionation of the supernatant and the sludge granular portion after the anaerobic/aerobic phases.

a   b

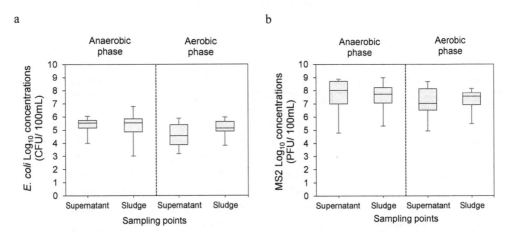

Figure 3.4 Log₁₀ concentrations of E. coli bacteria (a) and MS2 bacteriophages (b) in the supernatant and granular fractions at the end of the anaerobic and aerobic operational phases.

Table 3.1 shows the *p*-values obtained from the comparison of the phases. At the end of the anaerobic plug flow feeding, the median of the *E. coli* concentration for both supernatant and the sludge fraction was 3.8×10^5 CFU/100 mL (or 5.5 Log_{10}) (Figure 3.4, a). At the end of the aerobic phase (110 minutes of aeration), the *E. coli* median in the supernatant was reduced to 3.6×10^4 CFU/ 100 mL, or 4.6 Log_{10}, while the sludge fraction kept a more or less similar median concentration of 1.4×10^5 CFU/ 100 mL, or 5.1 Log_{10}). Both reductions were significantly different ($p \ll 0.05$, Table 3.1) than in the previous anaerobic phase. For the MS2 bacteriophage (Figure 3.4, b), a median concentration of 1.0×10^8 PFU/100 mL (or 8.0 Log_{10}) was measured at the end of the anaerobic plug flow feeding in the supernatant. The median of the counts in the sludge fraction was 5.4×10^7 PFU/ 100 mL (or 7.7 Log_{10}). After the aeration phase, a significant difference ($p = 0.02$, Table 3.1) was observed in the supernatant portion with 1 Log_{10} unit reduction in the median counts (1.1×107 PFU/ 100 mL or 7.0 Log_{10}). In the sludge fraction, no significant differences ($p > 0.05$, Table 3.1) were observed when comparing the median of the aerobic phase (3.7×10^7 PFU/100 mL, or 7.6 Log_{10}) with the median of the previous anaerobic phase.

Table 3.1 p-values obtained from the Wilcox test comparing concentrations observed in the supernatant and the sludge fraction after each anaerobic and aerobic operational phase.

Organism	Sampling point	n	W	*p*-value
E. coli	Supernatant Anaerobic-Aerobic phase	13	90	*0.0005*
	Granules Anaerobic-Aerobic phase	13	83	*0.0061*
MS2 bacteriophage	Supernatant Anaerobic-Aerobic phase	9	42	*0.0195*
	Granules Anaerobic-Aerobic phase	8	29	0.4961

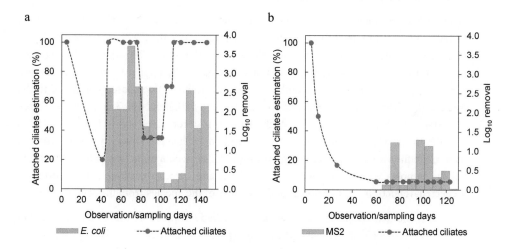

Figure 3.5 Attached ciliates abundance estimation (%) calculated based on the microscopic observations compared with the removals of E. coli bacteria (a) and MS2 bacteriophage (b).

3.3.4 Relationship of attached protozoa and surrogates' removal

Higher organisms typically present in the crushed granular sludge inoculum of the reactor were not microscopically observed when both reactors (AGS_*E. coli* and AGS_MS2) achieved their stable operation before adding *E. coli* and MS2. The stability was indicated by well-shaped granules that were formed and the accomplished stable COD and PO_4-P removal - after approximately two months. As soon as the reactors were inoculated with *E. coli* or MS2 bacteriophages, a sudden bloom of stalked ciliated protozoa attached to the granular surface was observed in the AGS_*E. coli* reactor, but not in the AGS_MS2 system. The attached ciliates occurrence in the granular samples was determined and compared with the removal of *E. coli* and MS2 (Figure 3.5). In the AGS_*E. coli* reactor, high activity (abundance and mobility) of attached ciliates was observed on days 48, 76, 113, and after 120. However, a decrease in the stalked ciliated activity occurred between days 86 and 105, which also coincided with both the reduction of the NH_4-N concentration shown in Figure 3.2, c and a reduction of the *E. coli* removal in the system (Figure 3.5, a). In contrast, in the AGS_MS2 reactor, no massive changes were observed for the stalked ciliated community when the system was spiked with the MS2 bacteriophage, coinciding with lower removals during the studied period.

3.3.5 Eukaryotic community analysis

The eukaryotic microbial community was characterised by 18S rRNA gene analysis (Figure 3.6 and Figure 3.7). The analysis covered more than 99% sequencing depths (Table 3.2), which is sufficient to cover the whole community. The index used to estimate the number of the species (abundance) in a community belonging to individual classes, Chao1 [110], showed that the number of species (richness) decreased compared to the seed sludge. That is from 818.3 in the seed sludge to 589.5 and 493.3 in the AGS_*E. coli* day 104 and day 160, respectively. At the same time, it remained almost the same (818.4) for the AGS_MS2 reactor (day 90). Instead, the Shannon diversity index used to determine the variation of living organisms [111], showed a reduction in the diversity in all the samples with values of 5.39, 2.03, 3.01, and 2.96, respectively. It confirms, an expected, reduction of the richness and evenness of the species from the seed sludge.

The most abundant (top 10) species in all the studied samples are described at the phylum levels (Figure 3.6) of the eukaryotic phylogenetic classifications. Mostly free-living organisms such as nematodes, tardigrades, and rotifers were likely removed via the effluent along with other particulate and suspended solids. Ascomycota and unidentified eukaryote were the most abundant groups found in the laboratory-scale systems.

Table 3.2. Alpha diversity indices for the samples based on the 18S rRNA gene libraries

Samples ID	Sequences	OTUs	Chao1 [a]	Shannon [b]	Simpson [c]	Good's coverage [d]
Seed sludge	123062	58050	818.3	5.394	0.932	0.997
AGS_*E. coli* (day 104)	109219	58094	589.5	2.028	0.603	0.997
AGS_*E. coli* (day 160)	98338	58082	493.3	3.058	0.682	0.998
AGS_MS2 (day 90)	147912	58082	818.4	2.955	0.691	0.996

[a] Chao1 abundance of individuals belonging to a certain class in a sample.
[b, c] Shannon and Simpson diversity index, the higher the number the greater the diversity.
[d] Good's coverage, a higher number represents more sufficient of sequencing depths.

Figure 3.7 summarises the most abundant protozoa phyla at their class and genus level. The most abundant genera in the seed sludge were *Rhogostoma* (18.9%); followed by *Telotrochidium* (12.4%), *Opisthonecta* (4.8%) and *Epistylis* (2.6%) which belongs the Oligohymenophorea class. Other peritrich ciliates genus such as *Pseudovorticella*, *Vorticella* and *Vorticellides* were in abundance between 0.1 and 0.7%. Most of the target genera were reduced over time in the laboratory-scale samples, i.e., *Telotrochidium* to 2.7% and *Epistylis* to 0.2 % in the sample taken in day 104 - which was during the likely copper contamination in the AGS_*E. coli* reactor. Notably, in day 160, when the AGS_*E. coli* system was again stable, *Telotrochidium* highly recovered to 23.2% and *Epistylis* to 1.8%. Regarding the genus *Rhogostoma,* irrespective the circumstances, the genus was prevalent and highly abundant in the AGS_*E. coli* reactor. It appeared to be slightly affected in day 104 (46.2 %) for the undesired copper addition [112], but fully recovered in day 160 (56.5%). Regarding the AGS_MS2 reactor sample in day 90, apart from the genus *Rhogostoma* (with a relative abundance of 29%), *Telotrochidium* (1.7%), and *Epistylis* to 0.2% were as conventional as the AGS_ *E. coli* in day 104. For the rest of the community their relative abundance was between 0.01 and 0.25%, but more diverse than the AGS_ *E. coli* reactor.

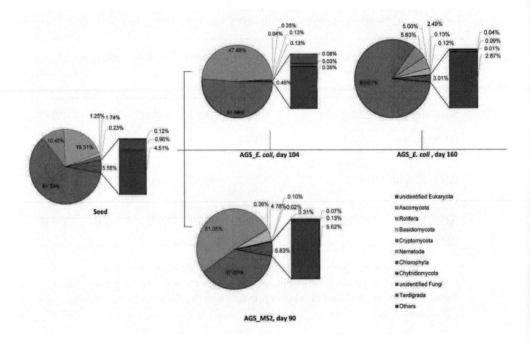

Figure 3.6 The relative abundance of the ten more frequent genome sequences at the phylum-level (a) and genus-level (a) taxonomy for the seed sludge, AGS_E. coli and AGS_MS2 reactor samples.

3.3.6 Predation recorded using fluorescent staining

Figure 3.8 shows pictures obtained when recording the ingestion of the *E. coli* by attached ciliated protozoa using dsGreen labelled *E. coli* bacteria. The pictures confirmed *E. coli* predation by stalked ciliates. The stained bacteria were visible inside the vacuoles of the organisms, and the *E. coli* bacteria were also visible embedded in the granular sludge matrix. Upon ingestion, the bacteria were concentrated, but the technique was not suitable for quantification of the partitioning of *E. coli* bacteria between supernatant fraction, attached to the granular surface or inside of the granular biomass.

Class	Genus	Seed	AGS_E. coli (day 104)	AGS_E. coli (day 160)	AGS_MS2 (day 90)
Mediophyceae	*Stephanodiscus*	0.00%	0.01%	0.01%	0.04%
Craspedida	*Monosiga*	0.02%	0.00%	0.00%	0.02%
	Salpingoeca	0.02%	0.00%	0.00%	0.01%
Hyphochytriomycetes	*Rhizidiomyces*	0.37%	0.01%	0.00%	0.04%
Ichthyophonae	*LKM51*	0.02%	0.01%	0.02%	0.06%
Labyrinthulomycetes	*Sorodiplophrys*	0.05%	0.02%	0.05%	0.56%
	unidentified_Labyrinthulomycetes	0.01%	0.00%	0.00%	0.05%
Litostomatea	*Entodinium*	0.00%	0.00%	0.27%	0.02%
	Acineria	0.00%	0.00%	0.00%	0.03%
Oligohymenophorea	*Telotrochidium*	12.39%	2.72%	23.24%	1.70%
	Epistylis	2.61%	0.17%	1.75%	0.20%
	Opisthonecta	4.80%	0.49%	0.09%	0.25%
	Pseudovorticella	0.68%	0.12%	0.10%	0.15%
	Vorticella	0.55%	0.04%	0.17%	0.07%
	Colpidium	0.00%	0.04%	0.06%	0.09%
	Paramecium	0.04%	0.02%	0.02%	0.18%
	unidentified_Conthreep	0.50%	0.01%	0.03%	0.02%
Peronosporomycetes	*unidentified_Peronosporomycetes*	1.20%	0.03%	0.12%	2.13%
	Pythium	0.05%	0.01%	0.06%	0.18%
Phyllopharyngea	*Tokophrya*	1.99%	0.02%	0.03%	0.08%
	unidentified_Conthreep	0.27%	0.02%	0.05%	0.02%
	Discophrya	0.01%	0.00%	0.00%	0.04%
	Acineta	0.00%	0.00%	0.01%	0.02%
	Trochilia	0.00%	0.01%	0.00%	0.01%
Prostomatea	*Prorodon*	0.23%	0.02%	0.01%	0.17%
Dinophyceae	*Scrippsiella*	0.00%	0.01%	0.00%	0.00%
Spirotrichea	*Aspidisca*	1.50%	0.09%	0.01%	0.12%
	Rimostrombidium	0.00%	0.000%	0.00%	0.00%
	Euplotes	0.00%	0.000%	0.00%	0.00%
Ichthyophonae	*Anurofeca*	0.01%	0.000%	0.01%	0.01%
unidentified_Eukaryota	*Rhogostoma*	18.87%	46.194%	56.50%	29.03%
	Blastocystis	0.01%	0.01%	0.01%	0.01%
Armophorea	*Metopus*	0.002%	0.00%	0.00%	0.00%
Plagiopylea	*Trimyema*	0.00%	0.000%	0.002%	0.002%

0% 100%

Figure 3.7 Heat map at the genus-level for AGS_E. coli and AGS_MS2. Genera comprised of the most abundant protozoa phyla.

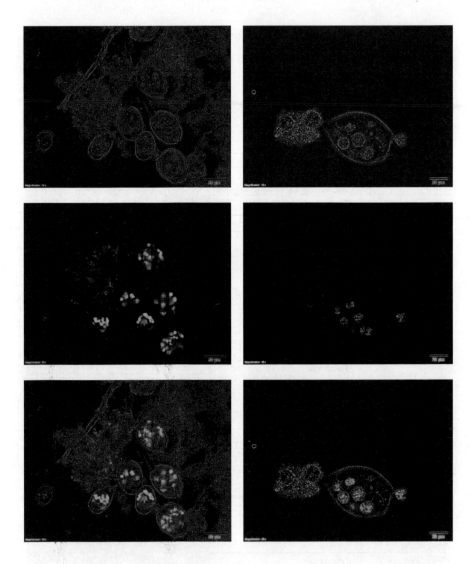

Figure 3.8 Phase-contrast picture (top), fluorescence microscopy (dsGreen) (middle) and their overlap (bottom).

It shows the ingestion of the bacteria by a colony of stalked ciliates (left side) and a single stalked ciliate (right side). The bar represents 50 and 20μm, respectively.

3.3.7 Attachment kinetics

Attachment tests were carried out with the surrogates (*E. coli* bacteria and MS2 bacteriophages). The attachment kinetics are represented in Figure 3.9. Two concentrations were tested versus the exposure time. Regardless of the concentration and the target organism, a speedy attachment occurred onto the granules in the first 15 min (Figure 3.9, a and Figure 3.9, b). Moreover, as can be observed in Figure 3.9, c and Figure 3.9, d, sharper curves can be observed for the lower concentrations of both organisms, meaning that a relative faster attachment occurred when compared with the more concentrated samples.

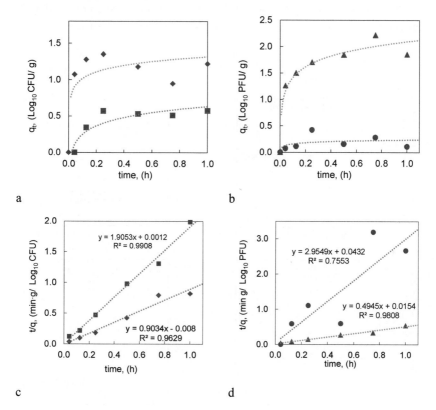

a   b

c   d

Figure 3.9 Attachment kinetics for: a) E. coli and b) MS2 bacteriophages. Linearized pseudo-second order kinetics for c) E. coli and d) MS2 bacteriophages.

Fitted lines and equations are shown for each data set. Initial concentrations: E. coli 1.3x10^7 CFU/ 100 mL (■) and 2.5x10^9 CFU/ 100 mL (♦); MS2 bacteriophages 1.6x10^6 PFU/ 100 mL (●) and 5.5x10^9 PFU/ 100 mL (▲).

3.3.8 Contribution of settling in the AGS reactors to the removal of the faecal surrogates

Additional tests were carried out to better understand the surrogates' removal in the liquid phase shown in the curve of Figure 3.3, b, specifically from the point measured in the supernatant at the end of the aeration to the mixed effluent. For *E. coli* bacteria, it seems settling responsible for a 0.5 Log_{10} removal; while for the MS2 bacteriophage, an increase of 0.1 Log_{10} is measured. Results from the settling test (Figure 3.10) shows that regardless of the initial concentrations, ($2.0x10^5$ and $5.3x10^7$ CFU/100 mL for *E. coli*, and $2.4x10^3$ and $4.7x10^9$ PFU/100 mL for MS2), no differences were measured in the liquid bulk right after settling, nor in the liquid fraction of the treated effluent. Therefore, settling forces were discarded as factors to explain the variations in the long-term reactors.

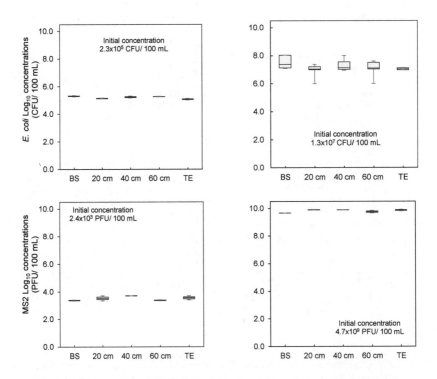

Figure 3.10 E. coli and MS2 counts obtained from the settling batch tests.

Concentrations correspond to the medians of liquid bulk before settling occurred (BS), at 20, 40 and 60 cm height from the discharge point and treated effluent (TE).

3.4 DISCUSSION

3.4.1 Reactor performance

The AGS reactors were both under operational conditions that triggered quick granular formation and an efficient reactor performance [113, 114]. Overall, both studied systems showed comparable performances to previous reports using similar substrates [115]. The addition to the *E. coli* and MS2 after day 47 for the AGS_*E. coli* reactor and day 69 for the AGS_MS2 reactor did not affect the general reactor performance. The measured water quality parameters shown in Figure 3.2 confirmed healthy systems with low effluent COD values, and high P-release after the anaerobic feeding of the AGS_*E. coli* and AGS_MS2 reactor, respectively. Except for the low NH_4-N removal in the AGS_*E. coli* reactor (from day 95 to 120), good conversion of NH_4-N to NO_2-N and NO_3-N during aeration was observed in both systems. The low NH_4-N removal in the AGS_*E. coli* reactor was most likely caused by the unexpected presence of copper in the system. Copper can be toxic for ammonia-oxidising bacteria [116] and other organisms such protozoa which can also interfere in the nitrification process [117]. After copper was excluded from the influent, the system recovered and showed a good performance until the end of the study.

3.4.2 Fate of the target surrogates during the operational conditions of the long-term reactors

The overall *E. coli* removal efficiency (2.2 Log_{10}) calculated for the AGS_*E. coli* reactor was within the range previously reported for *E. coli* in full-scale AGS systems. However, the MS2 bacteriophage removal (0.7 Log_{10}) was lower than reported [98]. It is worth mentioning that full-scale AGS systems contain both a large granular sludge fraction and a smaller flocculent sludge fraction [75]. Ali, Wang [21] has showed variances in the bacterial assembly depending on the different size of the aggregates including flocs, large and small granular fractions. Reactors here studied contained only granular sludge potentially impacting the removal of the faecal surrogates. Therefore, the fate of the two different faecal surrogates (*E. coli* and MS2 bacteriophages) focused on evaluating only the contribution of the granular fraction during different operational stages of two AGS systems to better understand their influence on the pathogen surrogates' removals.

3.4.3 Faecal surrogates' removals during the anaerobic plug flow feeding

In this study, an average reduction of 0.3 Log_{10} was observed in the liquid fraction for the faecal surrogates during the anaerobic stage in both AGS_*E. coli* and AGS_MS2 reactors. In the AGS systems, the influent is fed in a plug-flow mode, causing a concentration gradient from high to low in the water phase over the granular bed. A high concentration of the faecal surrogates present in the influent at the feeding point/bottom of the reactor was expected, which is diluted with the "clean" water at the top part of the reactor after feeding. Chong, Sen [118] and Vymazal [119] have shown that coliforms survive longer in anaerobic environments; thus, due to the fully anaerobic conditions, and the short time the feeding takes, the faecal surrogates' decay was negligible. Furthermore, the batch attachment tests showed that irrespective of the organisms, the *E. coli* bacteria and the MS2 bacteriophage quickly attach and saturate the granular surface when passing through the granular media (Figure 3.9). They kept an equilibrium concentration between the granules and the supernatant fraction as can also be observed in Figure 3.4. In case sludge waste occurs at this stage, as is practice in full-scale AGS treatment plants, a high concentration of surrogates (approximately 10^5 CFU/ 100 mL and 10^7 PFU/ 100 mL) would leave the system via the mixed liquor, a combination of the sludge and supernatant fraction here studied [120, 121]. Sludge treatment and dewatering will result in (additional) removal of pathogens [122, 123]. Therefore, based on previous findings, the minor reduction reported during the anaerobic period can be attained to a constant dilution effect of the influent with the remaining reactor media during steady state conditions.

3.4.4 Faecal surrogate removal during the aeration phase

During the aeration phase, the self-immobilised granular bed is continuously mixed, and granules are exposed to all the components remaining in the liquid bulk. Besides providing the right oxygen concentration in the systems [124], aeration provides high shear stress helping to form round-shaped granules [125]. This aerated phase is meaningful for some organisms that are oxygen depended such as protozoa [126].

Protozoa play a major role in wastewater treatment technologies [108, 127, 128]; they are unicellular, heterotrophs and eukaryotic organisms fed either by the absorption of dissolved nutrients or the ingestion of particulate matter including bacteria or organism present during the assimilation [129]. Pauli, et al. (2001) stated that protozoa could bind to bacterial conglomerates, such as flocs. In our laboratory-scale AGS reactors, the sludge beds were almost completely formed of granules. Smaller granules were quickly discharged during the effluent withdrawal inducing free-living metazoans such as nematodes, tardigrades, and rotifers to wash out (Figure 3.6).

At the beginning of this study, a reduction of the protozoa was measured using a microscope; however, once *E. coli* was added to the AGS_*E. coli* system, a bloom of stalked ciliated protozoa attached to the granular surface occurred. This tendency has been previously reported in laboratory-scale systems fed on particulate material [130, 131], and generally in full-scale wastewater treatment systems [92]. As can be seen in Figure 3.5, a, a higher *E. coli* removal was detected when a higher abundance of attached ciliated protozoa was microscopically observed. Ciliates are a dominant class in wastewater treatment systems [101, 132, 133]; they move through cilia and are subdivided into three categories, free-swimming, crawling and attached organisms. Examples of this are the free-swimming genera *Telotrochidium* and *Opisthonecta*, and the attached ciliates *Epistylis, Pseudovorticella, Vorticella* and *Vorticellides*, which were part of the Ciliophora phylum-level found in this study [134-136]. The free-living genus *Rhogostoma* from the Cercozoa phylum showed to be the most abundant in the AGS_E. *coli* reactor after both 104 (46%) and 160 (57%) operational days. It grows quickly under controlled (laboratory) conditions [137]. Öztoprak, Walden [138] describe this genus as a bacteria predator with a high diversity of clades, able to colonise a variety of habitats including wastewater matrices. Some species such as the *R. micra* are related to debris and bacteria [139], others (*R. epiphylla*) have been recognized as food selective with an affinity to predate yeast from the Ascomycota and Basidiomicota phyla [140]; organisms that were also present during steady conditions of the AGS_E. *coli* reactor (Figure 3.6).

Concerning the protozoa filter-feeding process called phagocytosis [141], the feeding starts by generating a water current, concentrating the particulate matter present in the liquid bulk while retaining the particles in size between 0.3 and 5 µm [74, 142], it includes our *E. coli* (1-2 µm). The process continues with the intake of the retained particulate matter in vacuoles. This intake was recorded in the individual batch tests when using fluorescently labelled *E. coli* (Figure 3.8). Phagocytosis was anticipated to occur only during aeration. Bacterivorous commonly found in anoxic environments such as *Metopus* and *Caenomorpha* from the Armophorea class were not present in the samples [143]. Interestedly, the anaerobic ciliates *Trimyema* (Plagiopylea) occurred but in a very low taxonomic abundance (0.002%) when the AGS_E. *coli* system was stable at day 160 [144]. Fenchel [126] stated that some other protozoa could also adapt and sustain their growth under oxygen limitations. It applies for *Euplotes* and *Rimostrombidium* from the Spirotrichea class. However, such genera were negligible in the studied systems. Matsunaga, Kubota [145] reported a greater diversity of uncultured eukaryotes, including phylogenetic affiliations found in this study. Overall, our results were consistent with eukaryotic molecular diversity studies using 18S rRNA gene analysis in different sewage systems.

For the MS2 bacteriophages, no significant changes were recorded when the bacteriophage was added to the reactor. Bacteriophages in general are very selective on their host [146]. The host specificity of MS2 bacteriophage used in this study depended mostly on *E. coli* F-pili [147, 148]. Therefore, infections of bacteria forming the granules (AOB, NOB and PAOs) were not expected and not measured based on the physicochemical reactor performance [149]. Therefore, due to lack of favourable conditions for reproduction of the MS2 bacteriophage, including a low host range of bacteria [150, 151], and slow infection cycles [152], cell lysis was assumed negligible.

Indeed, the MS2 bacteriophages addition did not induce the same stalked ciliated protozoa bloom rate as in the AGS_E. *coli* reactor. As a particle, MS2 bacteriophage (27 nm) is much smaller than the *E. coli* bacteria (1 by 2 μm) [153]. Besides the low rate bacteria erosion expected from the granules [113], and the lack of particulate material or any other bacteria in the synthetic influent, resulted in the reduced protozoa growth; indirectly affecting the MS2 bacteriophages removal. Overall, bacterivorous genera such as the free-swimming *Telotrochidium* (1.7%) and stalked *Epistylis* (0.2%) occurred. However, their relative abundance was lower compared with the values found for the AGS_*E. coli* reactor, which were 23% and 2%, respectively. The genus *Rhogostoma* (29%), which was also the most abundant organism found in the AGS_MS2 reactor, potentially grew by predating such free bacteria and fungi derived from the sludge granules, little contributing to the viral surrogates' removal. Deng, Krauss [104] reported that the free-living *Salpingoeca* (Craspedida) can use the phage as a potential carbon source by actively feeding on MS2. It coincides with the relative taxonomic occurrence of this flagellate in the AGS_MS2 reactor (0.01%) which was slightly lower than in the seed sludge (0.02%), but not abundant in the reactor fed only with *E. coli*. Hence, based on prey selection criteria, the protozoa feeding rate determined the grazing pressure on the added bacteriophages [154].

Regarding attachment, it was assumed that the granule surface reached an attachment equilibrium with the liquid bulk right after the anaerobic feeding. This assumption can be confirmed when looking at the long-term experiments in Figure 3.4, a, and Figure 3.4, b. The *E. coli* median concentrations of the granular fractions at the end of the aerobic phase kept the same order of magnitude than the previous phase, about 10^5 CFU/100 mL for *E. coli* and 10^7 PFU/ 100 mL for MS2. Therefore, recognising the role of the protozoa in our laboratory scale systems, the main removal mechanism during aeration can be documented as a one-phase process. On average 1 Log_{10} *E. coli* and 0.6 Log_{10} MS2 bacteriophage was measured during aeration by filter feeding protozoa [74, 99], which for the system fed only with MS2 were less abundant than in the AGS_*E. coli* reactor.

3.4.5 Contribution of the settling in the removal of the faecal surrogates

The settling batch tests executed in the additional column reactor showed that settling forces do not contribute to the removal of any of the studied faecal surrogates in the liquid phase (Figure 3.10). In the laboratory-scale reactors, the treated effluent is rapidly separated from the biomass due to their high density [155]. The surrogates initially attached to the granular surface settle along with the granules (Figure 3.4). Therefore, the effluent is a mixture of supernatant with high concentrations of suspended organisms and very small granules that did not settle during the short settling time (5 min). Such effluent composition explains the dynamic of the organism's depletion curve during settling time shown in Figure 3.3, b; which for the *E. coli* seemed that settling is adding to the overall removal. But for MS2 bacteriophage, similar concentrations between effluent and the liquid fraction after aeration were observed. Therefore, the variations found in the organism's depletion curve were based on the composition of the effluent samples, but not caused by any selection pressure.

3.4.6 Overall analysis

In this study, in the laboratory granular sludge system fed only with synthetic wastewater the removal of bacteria was higher than the removal of bacteriophages. The granules were saturated with high amounts of the surrogates (*E. coli* and MS2) achieving a saturation point during steady state. As previously mentioned in Section 3.4.3, such surrogates and actual pathogenic organisms will potentially abandon the system via waste sludge [73, 156]. In AGS full-scale systems, sludge waste normally occurs by a selection pressure that will discharge a high amount of flocculent sludge not commonly found in laboratory scale reactors [157]. Some physical properties such as cell mobility [158], opposite charge attraction [159], hydrophobicity [160, 161], and type of substrate added [22] might influence their attachment. *E. coli* and MS2 bacteriophages are considered good indicators of actual bacterial and viral pathogens in wastewater [61]. Their removal in full-scale AGS systems has been reported and compared with parallel CAS systems [98], along with the dynamics of antibiotic resistance genes [162].

Moreover, full-scale AGS systems are fed with complex substrates. Its characterization can have an impact on the eukaryotic structures diversity [163]. In our study, protozoa predation was the dominant mechanism to the actual removal of the surrogates. As protozoa are ideal grazers, their abundance and diversity could help to achieve better understand the pathogen removal mechanisms. A characterization of the protozoa community in a full-scale AGS systems is provided in Chapter 5.

3.5 CONCLUSIONS

Regardless of the organism *(E. coli* bacteria and MS2 bacteriophages), the bacterial and viral surrogates quickly attached to the granular surface saturating the granules during steady-state conditions. Therefore, physical removal plays a role when sludge waste occurs. During aeration, the *E. coli* bacteria and MS2 bacteriophages were reduced approximately 1 and 0.6 Log_{10}, respectively. Protozoa predation was the main contributor to the removal of *E. coli* during aeration. The 18S rRNA sequence analysis confirmed the occurrence of the genera *Pseudovorticella, Vorticella* and *Vorticellides*, which are attached ciliates from the phylum ciliophoran. A higher abundance of free-living genus *Rhogostoma* and the free-swimming ciliates *Telotrochidium* were also found. In the system fed with MS2 bacteriophages a similar eukaryotic community was observed, although at much lower amounts. Bacteriophages removal was low in the system spiked only with MS2. In full-scale AGS systems protozoa growth on the granular sludge fraction can significantly contribute to removing bacteria from the influent. The flocculent sludge fraction is responsible for further reduction in bacterial numbers and bacteriophages and needs more attention in future research.

4

CO-TREATMENT OF SYNTHETIC FAECAL SLUDGE AND WASTEWATER IN AN **AGS** SYSTEM

One of the treatment options of faecal sludge (FS) is co-treatment with domestic wastewater. As FS is far more concentrated than wastewater and variable in composition, co-treatment might negatively affect the biological treatment process. Implications of adding FS to AGS wastewater treatment systems was assessed by replacing 4 % (v/v) of the total influent flow of synthetic wastewater with synthetic FS in an AGS laboratory-scale reactor. Overall, with the FS addition the physicochemical performance of the reactor was impacted. It also resulted on a reduction of the granular size, an accumulation of solids in the reactor and a high portion of the granular bed became flocculent. A considerable protozoa *Vorticella spp* bloom attached to granules and the rest of the accumulated particles were observed; potentially contributing to the removal of the additional suspended solids which were part of the FS recipe.

This chapter is based on: *Barrios-Hernández, M.L., Buenaño-Vargas, C., García, H., Brdjanovic, D., van Loosdrecht, M.C.M. and Hooijmans, C.M. 2020. Effect of the co-treatment of synthetic faecal sludge and wastewater in an aerobic granular sludge system. Science of the Total Environment, 140480.*

4.1 INTRODUCTION

The provision of faecal sludge (FS) treatment is a necessary practice in urban areas not completely sewered where on-site sanitation is facilitated. The sludge generated on such facilities is highly concentrated in suspended solids, organic matter, and other contaminants; therefore, it needs a suitable treatment option to avoid the contamination of water resources and the risks to public health [164, 165]. According to Strauss and Montangero [166] and Lopez-Vazquez, Dangol [167], the FS can be classified based on the retention time of the sludge in the storage/collection container as either fresh, or digested sludge. Fresh sludge refers to the sludge that is disposed of, for instance, after being frequently emptied from non-sewered public toilets or bucket toilets (retention times of approximately several days up to a week). Digested sludge refers to sludge that is retained in the storage/collection container for several months or even years, and it has undergone a biochemical degradation process (e.g. sludge from septic tanks and/or pit latrines). Besides, the sludge can also be classified considering physicochemical characteristics as low, medium, and high strength FS [165, 168]. The biochemical degradation process that occurs in on-site systems depends on factors such as the temperature, retention time, presence of inhibiting substances, the water content of the sludge, among others [166, 168]. They may lead to the appearance of certain chemical compounds not present in the original FS; for instance, it is commonly observed in septic sludge the presence of different nitrogen species such as NH_4-N and NO_3-N as a consequence of the biochemical degradation of organic nitrogen [169].

Practices for FS treatment include landfill disposal, land treatment (agriculture practices), discharge in sludge treatment facilities (e.g. sludge drying beds, ponds, wetlands, anaerobic treatment systems, among others), and co-treatment with domestic wastewater [169]. Regarding the co-treatment with domestic wastewater, adverse effects on the performance of municipal WWTPs has been reported when the FS is added to such systems. For instance, Strauss, Larmie [170] and Heinss and Strauss [171], reported undesirable toxic effects both on the algae community in facultative ponds and on the methane-forming bacteria in anaerobic ponds due to the high concentration of ammonia present in the FS. Moreover, the co-treatment of FS on CAS-WWTPs, can severely compromise the quality of the treated effluent regarding the total suspended solids (TSS), the chemical oxygen demand (COD), and the nitrogen (N) and phosphorus (P) concentrations [165]. Higher oxygen demands in the biological reactor, odour issues, and the formation of scum and foam in the settling tanks can occur [171]. Furthermore, a modelling study carried out by Lopez-Vazquez, Dangol [167] confirmed that the effluent quality of CAS facilities not designed for co-treating FS can be very compromised. The authors recommended feeding small fractions of FS compared to the main influent wastewater flow (less than one per cent of fresh or digested FS in the municipal wastewater flowrate) to avoid disrupting the performance of the CAS system.

Regarding the AGS, the number of full-scale WWTP installations is increasing worldwide, both in regions exhibiting fully sewerage coverage as well as in regions provided with on-site sanitation systems. Pronk, Giesen [81] and Khan et al. (2015) reported on the performance of full-scale AGS-WWTPs co-treating septic sludge (Poland) and a high fraction of fresh FS (South Africa), respectively. The authors did not observe any significant adverse effects on the performance of the full-scale AGS systems regardless the sludge type; however, more detailed information is needed to properly conclude on the effects of co-treating faecal or septic sludge on the AGS systems. For instance, a complete influent characterization of the co-treated mixture reaching the AGS system was missing on such reports.

The preliminary information showed that AGS systems were not severely affected when co-treating septic sludge and FS; the desirable or maximum amount (fraction) of FS concerning the influent wastewater that can be appropriately treated by the AGS system is yet unknown. Moreover, no information has been reported on the potential implications for the treated effluent, the effect of the granular formation process and the system stability by the presence of such septic or FS fractions. This section, therefore, aimed at assessing the effects of co-treating FS on an AGS system. The performance of a laboratory-scale AGS system continuously fed with approximately 4 % (v/v) FS of the total influent wastewater was evaluated. The effects on the granule formation/stability and the presence of protozoa community in the AGS system were also assessed.

4.2 MATERIALS AND METHODS

4.2.1 Research design

Two AGS laboratory-scale systems were used, one reactor fed with synthetic municipal WW and a second reactor fed a mixture of synthetic FS and WW. The FS was introduced as 4% (v/v) of the WW flow, resulting in an influent with a much higher TSS, organic matter, and nutrient concentration than the influent of a control reactor. The influent was prepared by mixing both solutions (FS and WW) before entering the system. The effects of the FS addition on the reactor performance, the granulation process and morphology, and the occurrence of protozoa in the AGS system were analysed at different operational conditions in the reactor and compared to the control.

4.2.2 Development of the synthetic FS recipe

The synthetic FS developed in this study was based on synthetic recipes obtained from the literature for faeces and urine with high organic and nutrient fractions. Table 4.1 describes the urine and faeces composition used. The final synthetic FS recipe assumes that real FS is composed of 10% faeces, 9% urine, and 81% water.

Table 4.1. Urine composition (for 1 L of water) and measurements for the preparation of the synthetic faeces

Classification	Components for FS recipe	
Urine FS-1	NH_4NO_3 (g)	19.2
	$NaH_2PO_4 \cdot 2H_2O$ (g)	2.7
	KCl (g)	3.4
	$KHCO_3$ (g)	1.1
	Na_2SO_4 anhydrous (g)	2.3
	NaCl (g)	3.6
	HCl 32% (mL)	0.4
Urine FS-2	NH_4Cl (g)	4.0
	$Na_2HPO_4 \cdot 2H_2O$ (g)	3.8
Faeces FS-1 and FS-2	Cellulose (g)	24.1
	Psyllium husk (g)	21.1
	Yeast extract (10 - 12.5 w/w % Total N) (g)	36.1
	Miso paste (g)	21.1
	Olive oil (mL)	12.1
	NaCl (g)	2.4
	KCl (g)	2.4
	$CaCl_2 \cdot 2H_2O$ (g)	1.2
	Demineralised water (mL)	380

Faeces recipe

The synthetic faeces medium was prepared based on a recipe described by Penn, Ward [172]. The recipe was adapted to simulate a medium-strength synthetic FS. Cellulose, which is insoluble in water and slowly biodegradable, mimicked the particulate fraction of faeces. Psyllium husk was used as a source of dietary fibres and carbohydrates. Yeast extract and *E. coli* represented the bacterial content. Miso paste was used as a source of proteins, fats, fibres and minerals. Olive oil introduces the fat content (oleic acids) present in faeces. Other required inorganic minerals such as NaCl, KCl, and $CaCl_2$ were added as well. All the components were mixed and dissolved in demineralised water.

Synthetic urine

Two synthetic recipes for urine were used in this research. First, the solution was prepared as proposed by Udert and Wächter [173]. Hereafter, a new urine recipe was considered better reflecting fresh (rather than stored) urine, with a much lower nitrate concentration. The variation in the urine recipes resulted in two synthetic FS recipes, FS-1 and FS-2.

4.2.3 Reactors set-up

Two double wall column reactors were installed and operated as sequencing batch AGS reactors as described in Winkler, Bassin [174]. Both reactors had a diameter of 60 mm and a height of 1400 mm, with 2.9 L of working volume as described in Chapter 3, Figure 3.1. A bio-Controller (Applikon ADI1030, The Netherlands) and a bio-consoler (Applikon ADI1025, The Netherlands) automatically controlled the systems by continuously monitoring and adjusting the pH, dissolved oxygen, and temperature; the pH was adjusted to a value of 7.1 by dosing 1 M NaOH or 1 M HCl solutions to the bulk liquid of the AGS reactors during the aerobic phase. The temperature was set to 20°C. The air was recirculated by the aid of a recirculation pump (KNF, Germany) at an airflow rate of 6 L/ min; in such way, sufficient shear forces were provided to the reactors during the aerobic phase, while maintaining the DO concentration at the required set points of 1.8 to 4.2 mg/L. The effluent was discharged in the middle of the column (1.5 L); therefore, the exchange ratio was approximately 49% for both reactors.

4.2.4 Experimental procedures

The two AGS reactors were inoculated with granules from a full-scale AGS-WWTP located in Vroomshoop, the Netherlands. The initial TSS and volatile suspended solids (VSS) concentrations of the inoculum were 11.7 ± 2.6 g/ L and 9.5 ± 2.1 g/ L, respectively. The sludge volume index granules after 5 (SVI_5) and 30 minutes (SVI_{30}) was 40.1 mL/ g and 35.0 mL/ g, respectively. Sieved granules showed 67% of the particle size bigger than 0.22 mm, 30% between 1.0 to 2.2 mm and the rest smaller the 1.0 mm. Granules were crushed after characterisation and added to the AGS reactors.

One reactor was operated as a control for 215 days fed the synthetic WW - control reactor. The second reactor was operated for 209 days (AGS-FS reactor). Both systems were initially fed with the same synthetic WW (Phase I) and operated in 3-hour cycles as follows: 60 min anaerobic feeding, 110 min aeration, 5 min settling, and 5 min effluent discharge (Table 4.2). After 68 days of operation, the synthetic FS solution was added to the AGS-FS reactor as 4% (v/v) of the total synthetic WW influent flow (Phase II).

To ensure an optimal reactor performance, the 3 hours cycle was extended to 4 hours as follows: (i) an additional anaerobic stand-by period of 30 min was added (to enhance the hydrolysis of the slowly biodegradable particulates from the synthetic FS); (ii) a longer settling time of 10 min was introduced; and (iii) the aeration period was extended to 135 min. During Phase III, the AGS-FS reactor operational conditions were slightly adjusted to a settling time of 5 min, and to a longer aeration time of 140 min. The operational conditions on the control reactor were kept constant during the entire evaluated period.

Table 4.2. Operational conditions: Control and AGS-FS reactors

Operational conditions	Control reactor	Phases in the AGS-FS reactor		
		I	II	III
Feeding (min)	60	60	60	60
Stand-by period	0	0	30	30
Synthetic WW	✓	✓	✓	✓
E. coli medium			✓	✓
FS			✓	✓
Aerobic phase (min)	110	110	135	140
Dissolved oxygen (50%)	✓	✓	✓	
Dissolved oxygen (20%)	✓			✓
Settling (min)	5	5	10-5	5
Effluent discharge (min)	5	5	5	5
Number of cycles per day	8	8	6	6
Hydraulic retention time	5.8	5.8	6.8	7.7
Cycle length (h)	3	3	4	4
OLR (kg COD m^{-3}d^{-1})	2.9	2.9	7.2	8.0
NLR (kg N m^{-3}d^{-1})	0.4	0.4	0.5	0.3
PLR (kg PO$_3$-P m^{-3}d^{-1})	0.06	0.06	0.06	0.06
Operational days	215	67	29	113

4.2.5 Media composition

Both the control and the AGS-FS reactors were fed synthetic WW. The synthetic WW medium was prepared as described in Chapter 3, section 3.2.3. An *E. coli* solution was added to the reactor AGS-FS in Phases II and III; it was prepared by adding 25 mL of a stock solution (1 x 10^9 CFU 100/ mL of *E. coli* ATCC® 25992) to a vessel of 10 L of demineralised water. During Phase II, the AGS-FS reactor fed a mixed influent flow containing synthetic WW and the prepared synthetic FS-1 solution, and synthetic FS-2 solution during Phase III. Table 4.2 shows the loading rates added the system per phase.

4.2.6 Analytical determinations

Physicochemical analytical measurements

The FS recipes were characterised by measuring the total and soluble COD, nitrogen compounds, phosphate content, VSS and TSS immediately after preparation. Additionally, the nitrogen content was monitored in the FS recipes (FS-1 and FS-2) after three days of preparation to determine their variability. For the reactors, their performance was evaluated by weekly monitoring standard water quality parameters such as soluble COD, nitrogen compounds, and phosphate in samples taken from the influent, after the anaerobic feeding, after the anaerobic stand-by period and effluent of the AGS-FS and the control reactor. The following treated effluent standards were considered as reference for a good reactor performance: COD < 125 mg/ L, the sum of the NH_4-N, NO_3-N and NO_2-N < 15 mg/ L and PO_4-P < 2 mg/ L [33]. The COD measurement was performed according to the Closed Reflux-Colorimetric Standard Method [76]. NH_4-N, NO_3-N, NO_2-N and PO_4-P were measured using LCK cuvette tests (manufacturer: Hach®). Samples were filtered through a 0.45μm Millipore™ membrane filters previous the analysis. TSS and VSS were determined according to the Standard Methods for sludge and effluent samples [76]. The solids retention time (SRT) was calculated considering TSS concentrations in both the reactor and the treated effluent. The granular bed was weekly measured using the column volume scale previously implemented in the glass columns.

Sludge settle-ability and (granular) size distribution

The sludge settle-ability was determined by the SVI_5 as in Pronk, Abbas [22]. Samples were taken on days 69, 96, and 150 from the control reactor; while on day 57 (Phase I), 102 (Phase II) and 144 (Phase III) from the AGS-FS reactor. On day 69 and 150, the sludge was separated using a sieve with a particle size of 212 μm to determine the granular (> 212 μm) and the flocculent (< 212 μm) fractions. To determine the average granular size distribution, additional sludge samples were taken on day 118 from the control reactor and from the AGS-FS reactor. The size of the granules was measured by taken images with an Olympus SC50 camera adapted to an optical microscope; the images were later analysed with the ImageJ software [175]. The information was processed using statistical distributions and presented in histograms.

Granular morphology and structural composition of the granules

The morphology and microstructural composition of the granules were observed using a scanning electron microscope (SEM). Analyses were carried out in granular samples taken on the operational days 10, 64 and 103 of the AGS-FS reactor; and on day 110 of the control reactor. For the SEM determinations, the samples were fixated with a 0.5% formaldehyde and 0.5% glutaraldehyde solution after rinsing the samples three times with a phosphate buffer solution (1M K_2HPO_4 and 1M KH_2PO_4) at pH 7.2. After providing enough fixation time (16 hours), the samples were dehydrated using first an ethanol/water solution gradually increasing the ethanol/water ratio as follows: 30, 50, 70, 80, 90, and 100%. The solution was replaced three times at each ratio every ten minutes. Later, the same procedure was performed but using acetone instead of ethanol. The dried samples were finally placed on stubs using double-sided carbon adhesives double-coated with a gold sputter-coater before applying imaging in a JSM-6610 SEM at an acceleration voltage ranged from 5 to 10 kV.

4.2.7 Data analysis

Shapiro-Wilk normality test was applied to the measuring water quality parameters. Since not all the tested data were normally distributed, the non-parametric Wilcoxon rank-sum test (W) was used to compare the mean rank between the measurements taken from the influent and effluent samples both after the addition of the FS recipe (Phases II and III). A p-value ≤ 0.5 was used to indicate significance. Results were computed (1) to assess whether the measured parameters were significantly different from the media (substrate concentration variability); (2) to determine whether the two FS recipes were significantly different from each other, and (2) to evaluate whether the performance based on effluent concentrations of the AGS system differed when comparing the two studied phases.

Moreover, the Spearman's rank correlation ($\alpha = 5\%$) was applied to the solid concentrations in the reactor vs effluent to determine the relationship between biomass grow/accumulation in the reactor and the concentration measured in the effluent. The analyses were performed using R Core Team [80].

4.3 RESULTS

4.3.1 Characterisation of the medium-strength synthetic FS

Table 4.3 shows the average characterisation achieved for the FS recipe and the composition of the combined synthetic WW+FS. Apart from the PO_3-P concentrations, the Wilcoxon rank-sum test executed to the average COD and NH_4-N measurements showed a significant difference ($p < 0.05$) during Phase II and Phase III, representing two different recipes. Their addition to the AGS reactor resulted in fluctuations in the COD, nitrogen and PO_3-P content of the combined influent (Figure 4.1, Phase II and Phase III time 0 min). However, except for the nitrogen content, the concentrations that were weekly measured were not significantly different from the median value ($p > 0.05$). The separated degradation test performed to the FS recipes after three days of preparation confirmed that the NO_3-N concentration increased after three days from 14.9 to 22.2 mg NO_3-N/ L and 16.0 to 20.5 mg NO_3-N/ L in the FS-1 and FS2, respectively. This effect confirmed that the variations in the mixed influent concentrations were a result of the quick degradation of the FS recipe.

Table 4.3. Chemical characterisation of municipal synthetic WW, FS synthetic recipe, and mixed influent solutions.

Parameter	Concentrations (mg/ L)			Concentrations of combined influent (mg/ L)	
	WW	FS-1	FS-2	WW+FS-1	WW+FS-2
TSS	0	12,180	9,702	307	307
VSS	0	10,910	9,468	250	250
COD total	403	24,300	29,740	1170	1477
COD soluble	403	11,450	9,960	795	1151
NH_4-N	60	325	277	86	62
NO_3-N	0.0	300	18	10.8	1.3
NO_2-N	0.0	0.04	0.02	12.4	0.1
PO_4-P	8.5	93.8	96.7	13	12

Table 4.4. p-values of the comparison between the component of the FS recipe added in Phase II and Phase III in the mixed influent.

It includes data from the performance of the reactor based on effluent concentrations. The starts (*) flag the levels of significance, "ns" = not significant. Number of samples 23.

Parameter	Mixed influent			Treated effluent		
	W	p-value		W	p-value	
COD	23	0.013	*	83	0.07	ns
N	126	< 0.0001	****	126	< 0.0001	****
PO$_3$-P	35	0.66	Ns	88	0.12	ns

4.3.2 Evaluating the continuous performance of the AGS reactors

Figure 4.1 shows the average COD, PO$_4$-P and nitrogen concentrations for the control and AGS-FS reactor. The results are presented at different stages of the cycles of the AGS reactor as follows: (i) influent (at 0 min); (ii) after anaerobic feeding (62 min); (iii) after anaerobic stand-by (92 min); and (iv) treated effluent (180 min for the control, and 240 min for Phase II and III in the AGS-FS reactor). Additionally, Figure 4.2 shows the cycle's profiles for the AGS-FS reactor after the addition of the FS. Weekly measurements are shown in Annex 2, and Figure 4.9. In the control reactor, on average, 80% of the COD was consumed during the anaerobic feeding (from 376 to 77 mg COD/ L). The average consumption rate during this the anaerobic feeding was 86.8 mg COD g/ VSS·h. The rest (up to 90% of the COD) was consumed during aeration. It resulted in an effluent concentration of 38 mg COD/ L, corresponding mostly to the EDTA added in the recipe (which is non-biodegradable).

Regarding the PO$_4$-P, removal of more than 90% was observed starting on day 45 and onwards. The PAOs community seemed to be present and active releasing PO$_4$-P in the anaerobic phase of the cycles at concentrations ranging from 22 mg PO$_4$-P/ L to 67 mg PO$_4$-P/ L on operational days 27 and 67, respectively. The net PO$_4$-P uptake during the aeration phase of the cycle was 6.0 mg P g/ VSS·h), resulting in an average effluent concentration of 0.87 mg PO$_4$-P/ L.

Regarding the nitrogen removal performance of the reactor, the ammonium-nitrogen concentrations in the influent (65 mg NH$_4$-N / L) was reduced by approximately half (32 mg NH$_4$-N/ L) after one-hour feeding (62 min) in the anaerobic phase mostly due to dilution effects of the cycle. During the aeration cycle, an average removal rate for ammonium-nitrogen of 2.5 mg NH$_4$-N g VSS·h was reported (after operational day 47 and onwards). When starting up the control reactor, ammonium-nitrogen effluent concentrations were ranging from approximately 15 to 41 mg NH$_4$-N/ L. Complete nitrification was observed after operating the reactor for approximately 102 days; ammonium-nitrogen concentrations lower than 0.7 mg NH$_4$-N/ L were measured. The average nitrogen measured in the effluent was 22 mg/ L for the evaluated period. The ammonium-nitrogen was almost completely nitrified and the total nitrogen observed in the effluent corresponded mostly to NO$_2$-N and NO$_3$-N indicating that full denitrification was not achieved. During the operation of the reactor, the dissolved oxygen saturation was reduced from 50% to 20% (as indicated in Table 4.2) and improvements on the NO$_2$-N and NO$_3$-N effluent values were observed from 16 to 5.4 mg NO$_2$-N/ L and from 5.7 to 1.4 mg NO$_3$-N/ L. The AGS-FS reactor during Phase I (66 operational days) showed similar trends as the control reactor.

In Phase II of the AGS-FS reactor, starting on operational day 67, the carbon source increased to an average value of 795 mg COD/ L in the anaerobic feeding phase. As can be observed in Figure 4.1, a, the operational changes introduced during this studied period (i.e. an extra anaerobic stand by period of 30 minutes) resulted in an average reduction of the organic matter concentration from 320 mg COD/ L at 62 min (end of anaerobic phase) to 274 mg COD/ L at 92 min (end of stand-by period). The specific anaerobic consumption rate (including the stand-by period) of this phase was 35.4 mg COD g VSS·h. Contrary to Phase I, at the operational day 81 showed a potential lower PAOs activity since not all the acetate was taken up; concentrations of acetate of 128 mg Ac/ L, 50 mg Ac/ L, and 1.6 mg Ac/ L were reported at 62 min, 92 min, and in the effluent, respectively.

The P-uptake ratio was also reduced compared to the performance of the reactor during Phase I, from 6.0 to 2.0 mg P g VSS·h. P-release values during the anaerobic feeding (62 min) from 24 to 48 mg PO$_4$-P/ L (average of 41 mg PO$_4$-P/ L) were measured; lower than the P-release values in Phase I (from 22 to 67 mg PO$_4$-P/ L). No additional activity was observed by the PAOs during the extra (added) anaerobic period. The observed concentration in the treated effluent was on average 0.7 mg PO$_4$-P/ L.

On the subject of the nitrogen removal, the average ammonium-nitrogen removal rate dropped from 2.5 (Phase I) to 1.2 mg NH$_4$-N g VSS^{-1} h^{-1}; consequently, the average ammonium-nitrogen concentration in the treated effluent increased up to 19 mg NH$_4$-N/ L. In addition, the average nitrite concentrations increased from 7.6 to 15 mg NO$_2$-N/ L, while nitrate went from 4.1 to 5.2 mg NO$_3$-N/ L. However, it is important to highlight that the AGS-FS (Phase II) influent contains both NO$_2$-N and NO$_3$-N originated from the initial FS components (miso paste) and its potential decomposition. 39 mg Nitrogen/ L was measured in the treated effluent; although more than 50% was removed, nitrification and denitrification did not occur at full extent.

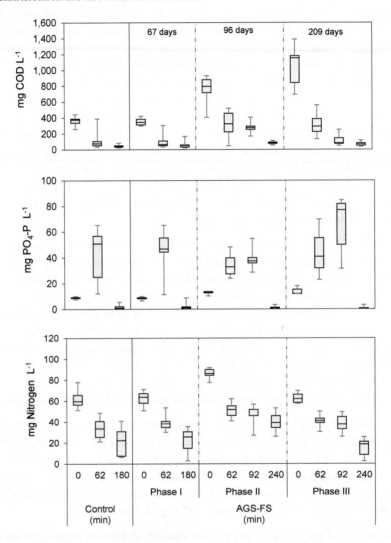

Figure 4.1. Average water quality performance of the control and the AGS-FS reactor at the different sampling points.

Influent (0 min), after anaerobic feeding (62 min), after anaerobic stand-by (92 min); and treated effluent (180 min for the control, and 240 min for Phase II and III in the AGS-FS reactor).

During Phase III of the operation of the AGS-FS reactor, the composition of the synthetic FS recipe was modified. The new FS recipe contained a higher organic content, resulting in a higher COD concentration (average 1152 mg COD/ L) in the AGS-FS during this phase (Figure 4.2, a). Moreover, nitrite (Figure 4.2, e) was not added to the new synthetic FS, resulting in a better P uptake performance during the anaerobic feeding phase. The average COD uptake during the anaerobic period in Phase III increased, the COD concentration went from 290 mg COD/ L at 62 min to 80 mg COD/ L after the anaerobic stand-by period (92 min). Moreover, an average COD concentration in the effluent of 73 mg/ L was reported, corresponding to a removal efficiency of above 96%. When comparing with the previous Phase II performance, COD effluent values were not significantly different (Figure 4.4).

An increase of the P-release (Figure 4.2, b) after the anaerobic feeding (62 min) was observed, the phosphate concentration went up to 41 mg PO_4-P/ L (compared to 31 mg PO_4-P/ L in Phase II). After the stand-by period (92 min) the average P-release was 77 mg PO_4-P/ L. After the operational day 160, the P-release reached a steady-state value of above approximately 80 mg PO_4-P/ L almost twice the value that was measured at 62 min. The P-uptake rate remained similar as in Phase II (2.0 mg P g VSS·h); however, the P concentration in the effluent was reduced to 0.3 mg PO_4-P/ L.

Regarding the performance on nitrogen removal during Phase III (Figure 4.2, c-e), the ammonium-nitrogen conversions slightly increased to 1.4 NH_4-N g VSS^{-1} h^{-1}, and much lower nitrate and nitrite concentration were measured compared to Phase II. Nitrate and nitrite concentrations of 1.3 mg NO_3-N/ L and 0.11 mg NO_2-N/ L were measured for the treated effluent, respectively. The system achieved concentrations lower than 15 mg Nitrogen/ L in the effluent after 186 days of operation. Thus, the lower nitrate concentration of the FS recipe seemed to have had a positive effect both on the nitrification and denitrification processes observed in the AGS-FS reactor.

Both reactors showed similar granular formation and steady bed volumes of approximately 400 mL. Figure 4.3, a, and b, show the TSS concentration in the reactors, as well as in the treated effluent.

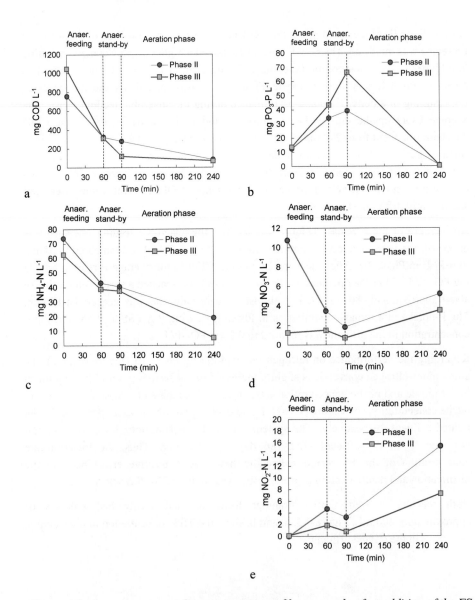

Figure 4.2. Average water quality parameters profiles per cycle after addition of the FS (Phase II and III) in the AGS-FS reactor. Suspended solids in the long-term reactor performance

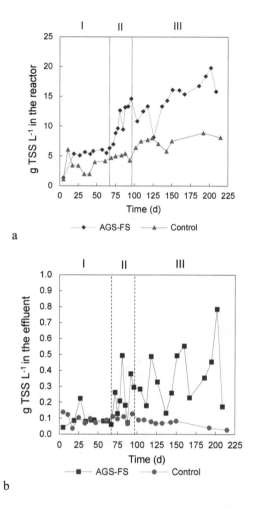

Figure 4.3. TSS concentration in the control and the AGS-FS reactors (a) and treated effluents (b).

The TSS concentration in the control reactor increased from 4 g TSS/ L to almost 9 g TSS/ L with an average VSS/TSS ratio of 0.81. The TSS concentration in the AGS-FS reactor also increased during the first 19 days of operation (from 1.2 to 5.3 g TSS/ L). During Phase I, the TSS concentration in the reactor reached a value of approximately 5.6 g TSS/ L with a VSS/TSS ratio of 0.81. The sludge bed in the AGS-FS reactor was comprised mostly of granules (> 95%) and the rest of flocs at this stage. An average TSS concentration in the treated effluent of 0.1 g TSS/ L was observed for both the control and the AGS-FS reactor (Phase I).

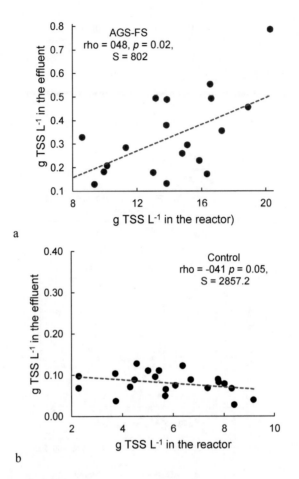

Figure 4.4. Relationship between the solids concentration in the effluent and the solids in the AGS-FS (a) and control reactor (b).

After the addition of the synthetic FS to the AGS-FS reactor (Phase II), the sludge bed volume increased mostly due to the higher concentration of suspended solids in the influent wastewater compared to the municipal synthetic WW. The concentration of solids in the influent was on average 307 ± 96 mg TSS/ L; i.e., 460 mg of solids added per cycle of the AGS reactor. The TSS concentration in the reactor increased steadily during Phase II as observed in Figure 4.3, a. The solids concentration in the reactor and the effluent showed some peaks and valleys, caused by extra sludge wastage via the effluent on top of the regular sludge waste (via taken weekly samples).

During the final stage (AGS-FS, Phase III), the TSS concentration in the influent remained approximately the same as during Phase II. The TSS concentrations in the reactor increased from 10.1 g TSS/ L on day 104 to 19.9 g TSS/ L on day 202. Low values were observed during day 125 and 200 that are attained to analytical measurements. The TSS concentration in the effluent increased from 0.1 g TSS/ L at the beginning of the Phase II to 0.3 g TSS/ L at the end of Phase III. There was a significant Spearman's rank ($p = 0.02$) correlation between the solids concentration in the reactor and the solids leaving the system via the effluent (Figure 4.4, a). Moreover, it was observed that the sludge bed in the AGS-FS reactor comprised of approximately 60% of granules and 40% of flocs at the end of Phase III. The dynamics of the granular and floating bed can be observed in Annex 2, Figure 4.10.

From the reactors' start-up to the operational day 83, the control reactor had an SRT value of 16 days; after day 83, the SRT value of the system was 20 days. The AGS-FS reactor (Phase I) operated on an average SRT value of 18 days. After the faecal sludge addition (Phase II), the SRT was reduced to 10 days. Due to the reactor operational changes, the SRT value reached 14 days at the end of Phase III.

4.3.3 Effects of FS on the granular settle-ability and size distribution

Both granular beds (control and AGS-FS reactors) exhibited a good settle-ability with SVI_5 values of 63 (at operational day 69) and 85 mL/ g (operational day 57), respectively. They showed the same SVI_5 value of 36 mL/ g on operational days 96 and 102, respectively. Hereafter, the SVI_5 in the control reactor decreased to 30 mL/ g (on operational day 150), and in the AGS-FS reactor (Phase III) increased up to 52 mL/ g on the operational day 144 (Phase III).

Regarding the particle size distributions, Figure 4.5 shows the granular size obtained from both the control and Phase III of the AGS-FS reactor on days 112 and 118, respectively. For the control reactor, an average granule diameter of 2.1 ± 0.5 mm was measured, and 95% of the granules were in a range between 1.2 and 3.0 mm. For the AGS-FS reactor, the size of the granules showed more variation. An average granule diameter of 2.1 ± 0.8 mm was measured and reported; however, only 69% of the granules were in a range between 1.2 and 3.0 mm. For 20% of the granules, the average diameter was reduced to 0.6 ± 0.1 mm, and the remaining 11% had an average diameter between 3.6 and 6.0 mm.

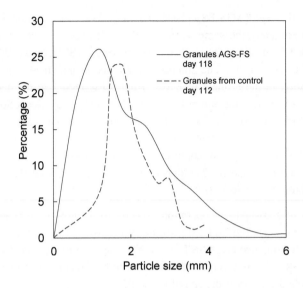

Figure 4.5 Particle size distribution of the granules in the control reactor and after the addition of the synthetic FS in the AGS-FS

4.3.4 Effects of the synthetic FS on the occurrence of protozoa

The protozoa community presence was evaluated in the control reactor and at the three different stages (Phases I, II, and III) of the AGS-FS reactor by SEM and optical observations. Both reactors (control and AGS-FS) were inoculated with 400 mL of crushed granular sludge and initially fed only municipal synthetic WW based on acetate as the carbon source.

Figure 4.6, a shows the SEM of a sludge sample taken from the AGS-FS reactor on operational day 10. The granules are under formation; therefore, on the surface of such granules, the SEM observations indicated the presence of agglomerations of cocci-shaped bacteria attained to stalks. A substantial reduction in the protozoa community population was observed; protozoa previously found in the inoculum (Figure 4.7, a and b) including rotifers and the typical genus of Peritrich ciliates subclass; i.e., *Vorticella* spp., *Carchesium* spp., and *Epistylis* spp., among others, were no longer present on the granular surface or in the liquid bulk. After operational day 64 of the AGS-FS reactor (Figure 4.6, b) and 110 days of the control reactor (Figure 4.6, d), systems were in a steady-state condition, and well-shaped granules were observed in the systems.

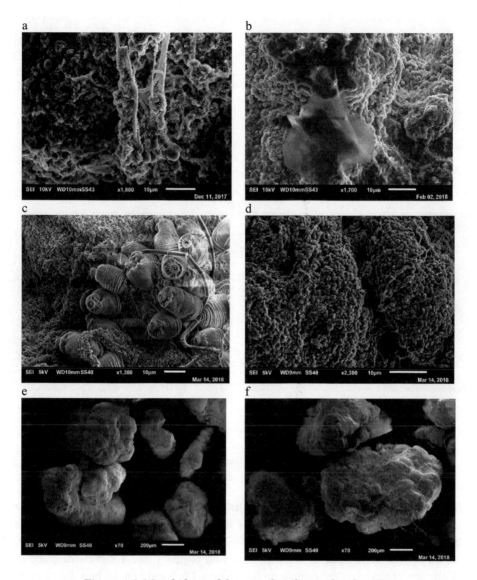

Figure 4.6. Morphology of the granules observed under SEM.

(a) After operational day 10 (Phase I of AGS-FS); (b) After operational day 64 (AGS-FS –Phase I); (c) After operational day 103 (AGS-FS reactors - Phase III); (d) After operational day 110 (control reactor); (e) and (f) Appearance of the formed granules from AGS-FS reactor and control after operational day 103, respectively.

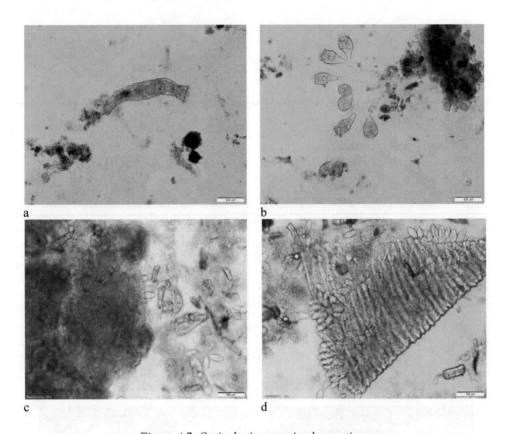

Figure 4.7. Optical microscopic observations.

(a) and (b) Higher organisms found in the inoculum; (c) Ciliates attached to the granular surface. (d) Ciliates grazing/attached on the psyllium husk particle.

SEM images showed a well-arranged granular surface fully formed of rod-shape bacteria lacking protozoa organisms in the granular surface of both samples. Interestingly, the inoculum used in the study was from a WWTP treating besides domestic wastewater, industrial WW discharged by a plastic recycling company and a particulate piece of plastic was observed on the granule surface (Figure 4.6, b and c shows how the granules formed of rod-shaped bacteria were fully colonised by stalked ciliates from the Peritricia subclass, mainly *Vorticella* spp. after the operational day 103 of the AGS-FS reactor (Phase III) [93]. At this operational stage, granules in the reactor were healthy but slightly smaller than the control (Figure 4.6, e and f). Additional optical microscopy (4x magnified) observations showed a condensed population of such organisms; they colonised both the granular surface (Figure 4.7, c), as well as most of the particle surfaces present in the FS recipe (Figure 4.7, d).

4.4 Discussion

4.4.1 Consideration for the development of a medium-strength synthetic FS

This study assessed the effect of co-treating two different synthetic FS recipes with synthetic municipal WW in a long-term operational AGS reactor. The use of the medium-strength FS recipe aimed at assisting countries with extensive on-site sanitation coverage (septic tanks) to better understand the dynamics of co-treatment of FS with wastewater in AGS systems [164, 176]. The synthetic FS was used to simulate the variety of biochemical components present in real FS such as bacteria debris, fibres, fats, minerals, proteins and carbohydrates. In regards to the urine simulant, initially, the solution used for FS-1 was developed based on the suggestions giving by Penn, Ward [172], and assuming that the urea was hydrolysed to ammonia, that 25% of the ammonia was already volatilised and that 50% of the remaining ammonia was oxidised to nitrate [173]. However, the final nitrate concentration was unrealistically high compared to real FS. Therefore, a second urine solution was developed and used in this study (FS-2) containing less nitrate.

The final FS simulant (FS-2) developed met the terms of the highly variable characteristics presented in Strande, Ronteltap [165] for septic tank sludge. Nevertheless, when using the FS recipe during the experiments, it was observed that the mixed influent exhibited high variability of COD and N components although it was kept at 4°C and replaced every three days. Changes might have been caused by the use of miso paste in the faeces simulant. This paste is made from soybeans, rice, salts, water and the filamentous fungus *Aspergillus oryzae*, which is rich in hydrolytic enzymes for the fermentation process [177], although *Lactobacillus* and *Bacillus* species may additionally be used [178]. Nout [179] stated that of all the nitrogen present in this type of products, half is present as amino-nitrogen. Because of the presence of easily available carbon and nitrogen, protein degradation and growth might have taken place in the mixed influent medium resulting in the high variability of COD and N components.

4.4.2 Effects of the FS on the continuous performance of the AGS reactor

For the AGS control reactor, a stable performance was observed during the entire operational time, and stable and mature granules were reported. The AGS-FS reactor also showed stable conditions and mature/stable granules during the operational conditions set for Phase I. The biological phosphorus removal observed in both reactors (control and AGS-FS- Phase I) agreed with previous AGS studies reported in literature using a similar influent P/Ac ratio [180, 181]. After the operational day 57 (for the control reactor) and

75 (for the AGS-FS reactor Phase I), acetate was fully consumed in both reactors as can be expected for a healthy acetate-fed system [182]. The resulting P-uptake of 6.6 and 6.0 mg $P/gVSS \cdot h$ for the control and the AGS-FS reactor was in agreement with the findings reported by Bassin, Kleerebezem [183].

Regarding the N removal, the initial reduction in the NH_4-N concentration at the end of the anaerobic feeding phase (after 62 min) in both reactors (control and AGS-FS Phase I) was due to a dilution effect and to a potential absorption of ammonium-nitrogen onto the surface of the granules [184]. The high concentrations of nitrite and nitrate observed after the aeration phase are due to an inefficient denitrification process observed both at the control and the AGS-FS reactors. The initial relatively high dissolved oxygen concentration set in the reactors (4.5 mg/ L) may have negatively impacted the denitrification process. A well-adjusted growth rate is necessary for anoxic conversions in the inner layers of the granule [174, 183]. Longer operational times could have been required to reach an optimal nitrification/denitrification rate in the systems. Therefore, the dissolved oxygen set-point was reduced in order to reach a concentration of 1.8 mg/ L in the reactor. Hereafter, lower nitrite and nitrate concentrations were measured in the effluent (Figure 4.1, d-e); however, complete denitrification was not observed during the entire operation of the reactors and high total nitrogen concentrations were always measured in the effluent. A better adjustment of the dissolved oxygen in the reactors may be needed to achieve complete denitrification.

After the synthetic FS addition, multiple effects on the system performance were observed. The extension of the anaerobic phase by an added stand-by period of 30 min seemed to facilitate the hydrolysis of the organic matter to more simple components (, a) [24, 130, 185]. However, it was observed that the increased anaerobic period was not long enough to ensure the complete organic matter uptake as occurred in the control reactor fed with synthetic WW. It seems that the PAOs were not able to store all the COD at the end of the anaerobic phase as can be seen by the high COD measured and the lower acetate consumption. One of the reasons was the high NO_3-N concentration present in the FS causing the anaerobic phase to be anoxic affecting the PAOs which could not sufficiently accumulate acetate or the hydrolysed substrate disrupting the P-uptake during the aeration phase [130, 181, 186]. Moreover, the increased denitrification affected the granular settle-ability and resulted in a higher solid wash-out via the effluent.

The SRT of the AGS-FS reactor decreased from 18 days in Phase I to 10 days in Phase II, which is unfavourable for slow-growing organisms such as ammonium or nitrite oxidisers [187]; consequently, less ammonium was oxidised during the aeration phase impacting on the expected simultaneous denitrification process. With the lower NO_3-N concentrations present in the FS synthetic recipe for Phase III, nitrite/nitrate no longer accumulated in the anaerobic feeding; hence, the PAOs activity improved by increasing the COD uptake/P release after the anaerobic phase as can be seen during 62 min in Figure

4.2 of Phase III compared to Phase II. Besides, due to operational changes and modification of the FS recipe, less biomass was wasted via the effluent, and the reduction of dissolved oxygen to 20% saturation enhanced the simultaneous nitrification/denitrification capacity of the system [188].

Moreover, as previously reported by Rocktäschel, Klarmann [189], the presence of the high concentration of TSS in the influent and its accumulation in the reactor led to high solids concentrations in the treated effluent. The TSS went from values lower than 0.01 g/ L for both the control and Phase I of the AGS-FS reactor to 0.8 g TSS/ L in the operational day 200 (phase III) for the AGS-FS reactor (Figure 4.3). The solid accumulation showed a significant relationship with the solids concentration in the effluent as well (Figure 4.4, a, and b). Even though such values are unusual in AGS full-scale WWTPs [75, 190], the results from our study showed implications for effluent quality that may require further attention. However, as mentioned by van Dijk, Pronk [96], the solids concentrations can be better studied in full-scale WWTPs where the sludge withdrawal take place in a separate process to enhance the effluent quality.

4.4.3 Effects of FS on granular formation and stability

Granules from the control and AGS-FS reactor Phase I showed a good settle-ability with an SVI_5 of 38 ml g^{-1}. This is slightly higher than the value reported by de Kreuk, Heijnen [30] for granules full-grown with acetate (SVI_5 24 mL/ g) but considerable lower than for flocs in CAS systems (100-150 mL/ g). The measured granular diameter (from 1.2 and 3.0 mm) was in line with the classification previously described by de Kreuk, Kishida [9] and Corsino, Di Trapani [191]. Moreover, the biomass yield measured in this study agreed with previous laboratory-scale studies confirming healthy granular systems using acetate as a substrate [189, 192].

In line with Cetin, Karakas [193] and Corsino, Di Trapani [191], the high amount of solids and the higher OLR as a result of the FS synthetic recipe in the influent had an impact on the particle size distribution of the granules. The size of 20% of the granules of the AGS-FS reactor was reduced to a range of 0.6 to 1.2 mm. However, the size remained within the reported standards for AGS systems [9]. Based on the water quality parameters, there were no noticeable adverse effects in case of the AGS-FS reactor. The average granular size was between the optimal size (0.7 − 1.9 mm) suggested by Zhou, Zhang [194] to enhance nitrogen removal efficiencies which was also shown in this study.

Moreover, a higher fraction of flocculent sludge (40%) was observed [192]. Similar effects were also reported by Liu and Tay [195], they encountered a detrimental effect on the granular fraction and faster growth of the flocs. They also found that the composition of the microbial community of the flocculent fraction hardly differed from the granular fraction when feeding a reactor with industrial wastewater (with concentrations ranging from 250 to 1800 mg COD/ L and 39 to 93 mg NH_4-N/ L). Selection pressure by reducing the settling time was as a measurement taken to enhance the granule fraction, which was also applied in this study (after initially increasing the settling time to 10 minutes, is was reduced again to 5 minutes) to maintain an optimal particle granular size.

4.4.4 Effects of the synthetic FS on the occurrence of protozoa

The inoculum used for seeding the AGS-FS reactor contained a variety of organisms (i.e., bacteria and eukaryotes), EPS, and other kinds of components such as (small) amounts of micro-plastics – originating from discharge to the sewer by a connected plastic recycling company. Rotifers and protozoa, which have been previously observed during the start-up of two AGS laboratory-scale reactors fed with particulate materials such as starch and maltose [130], were also observed. According to Zhang, Ji [196], it is assumed that the filaments and stalks of those higher organisms generate a support structure for the growth of the bacteria and the granule formation. Weber, Ludwig [94] explained the development of the granules in sequencing batch reactors systems (i.e., malthouse, brewery, and synthetic WW) in three phases. Firstly, protozoa get attached to both flocs and free-swimming particles developing a tree-like colony of ciliate stalks. Subsequently, bacteria colonise the stalks leading to the development of the granules. Later, some of the protozoa die or leave the biofilm through the treated effluent as free-swimming ciliates. In this study, during the start-up and maturation of the system, granules larger than 2.0 mm were observed in both reactors (control and AGS-FS); moreover, an agglomeration of bacteria was observed which was present in the inoculum as shown in Figure 4.7, a and, b; however, no protozoa were found on the surface of the granules or in the bulk liquid. They may have been washed-out during the effluent withdrawal. The remaining protozoa could have been embedded in the biofilm entirely covered by the same bacteria forming smooth and compact granules; therefore, protozoa were not seen under the light microscope.

From Phase II onwards, in line with Weber, Ludwig [94], de Kreuk, Kishida [130] and Corsino, Di Trapani [191] findings, an outgrowth of filamentous bacteria together with finger-type structures were expected to occur; this because of the addition of particulate material on the FS recipe. However, it was not the case in this study, a sudden bloom of protozoa occurred right after the addition of the synthetic, medium-strength FS. This present study validates Cetin, Karakas [193] findings, the addition of high solids concentrations in the systems will not always result in a filamentous organisms out-growth. Eventually, compounds such as microcrystalline cellulose, yeast extract, and psyllium husk could have induced the growth of the protozoa (which were most likely embedded in the granules). Unfortunately, no protozoa enumeration was done in this study for determining the growth and decay rate of these microorganisms in the reactors.

4.4.5 The relevance of the findings for future applications

The addition of synthetic FS to a healthy granular system helped to better understand the co-treatment dynamics in an AGS reactor. Our study showed that the AGS-FS system was able to treat synthetic FS being 4% of municipal synthetic WW total influent flowrate during 186 days of operation. Results confirmed that a functional AGS system can handle the co-treatment of FS with wastewater [81]. The good treatment performance was possible due to an extended duration of the anaerobic phase, a reduction on the dissolved oxygen set-point, and the changes in the settling time. Probably, those parameters are part of the operational conditions applied in full-scale AGS systems co-treating FS but are not yet reported. However, the operation of full-scale AGS systems will require attention when dealing with the high organic and nitrogen loads resulting from co-treating FS. Feeding conditions (i.e., FS dilution range and required pre-treatment) will need to be considered concerning the WWTP design capacity, as it was observed in this study that the additional particulate organic matter and the high nitrogen content coming from the FS impacted the AGS system performance. Furthermore, real FS has a high variation in composition that will require a proper characterisation before its addition into an AGS system. The behaviour of these different types of substrates (i.e., fresh or digested FS coming from pit latrines or septic tanks, among others) needs to be further evaluated.

4.5 CONCLUSIONS

The two adapted synthetic medium-strength FS (low and high NO_3-N) recipes well-represent digested medium-strength FS originating from septic tanks. The co-treatment of FS with synthetic wastewater required operational adjustments to prevent the deterioration of the effluent quality. The high NO_3-N concentrations present in the first FS recipe led to a disruption of the anaerobic conditions required for an optimal aerobic granular sludge system performance causing a floating sludge bed and solids wash-out. The second FS recipe with lower NO_3-N levels in the influent produced a better organic matter and nutrients removal. However, the addition of FS decreased the sludge settle-ability, and an accumulation of solids in the reactor occurred for both FS recipes. Moreover, due to the addition of particulate biodegradable organics, the average granular size was reduced, a higher fraction of flocculent sludge was perceived, and a sudden bloom of ciliates protozoa occurred. Further studies are necessary to determine the effect of real FS and its variabilities on full-scale AGS systems.

4.6 ANNEX 2

Table 4.5. p-values for the comparison of the variation of the influent concentrations with their expected concentration ($\alpha = 0.05$)

		Expected value	Mean	SD	Wilcox. Test	p-value	number sample
Phase II	COD	795	756	168	20	0.820	9
	N-N	108	86	5	93	**0.004**	9
	PO$_3$-P	13	13	1	16	0.779	9
Phase IIII	COD	1151	1048	227	39	0.395	14
	N-N	60	64	4	45	**0.011**	14
	PO$_3$-P	12	14	2	16	0.370	9

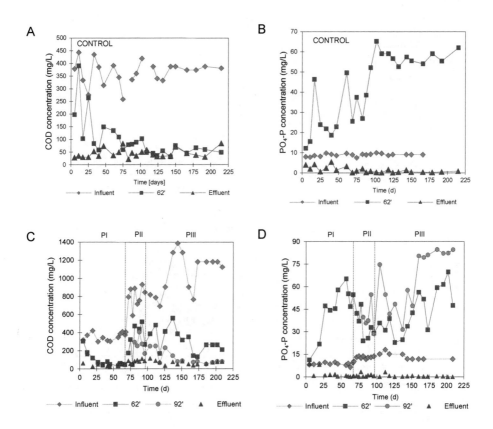

Figure 4.8. COD and PO₄-P profiles for the control reactor (A and B) and the AGS-FS (C and D) at different sampling points: influent, after anaerobic feeding (62'), after aeration (92') and effluent.

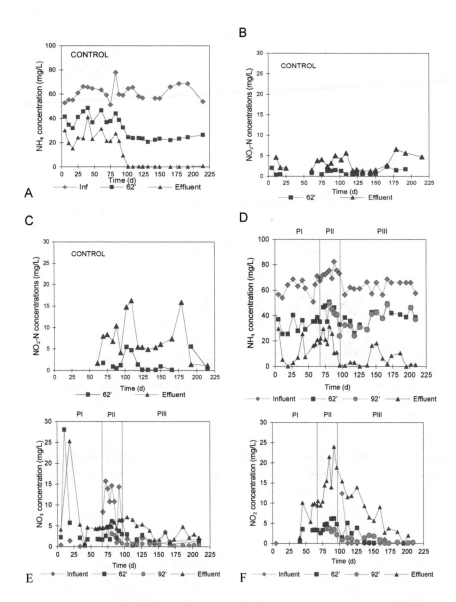

Figure 4.9 Chemical performance: (A) NH₄-N, (B) NO₃-N, (C) NO₂-N of the control reactor and (D) NH₄-N, (E) NO₃-N, (F) NO₂-N of the AGS-FS reactor at different sampling points: influent, after anaerobic feeding (62'), after aeration (92') and effluent.

Figure 4.10 AGS-FS reactor appearance before and after the FS addition.

A) Operational day 1, sludge inoculum; B) Operational day 34, granules already formed; C) Operational day 67; D and E) Operational day 70, AGS-FS reactor during feeding (5 min difference); F) Control column reactor.

5

EUKARYOTIC COMMUNITY CHARACTERISATION BY *18S rRNA GENE* ANALYSIS IN FULL-SCALE SYSTEMS

Eukaryotic structures (protists, fungi, rotifers) play a significant role in wastewater purification and disinfection. This study focusses on the determination of the diversity, occurrence and variation of the eukaryotic community in AGS systems using 18S rRNA gene sequencing analysis. Identification was done for wastewater influent, granular (mixed and fractionated into larger or smaller than 1.4 mm) and activated sludge samples of two parallel full-scale AGS and activated sludge systems treating the same influent. Results showed that the raw wastewater influent is more diverse in eukaryotes ($p \ll 0.05$) than the granular and activated sludge samples. Moreover, the taxonomic classification for the AGS systems showed a larger community of protist, fungi, chloroplastida, and animalia than the previously reported in literature using microscopic techniques. Protist structures *Colpidium, Opisthonecta, Vorticella, Peronosporomycetes, Tokophrya, Arcella, Gymnophrys* and *Rhogostoma* were the most abundant genera along with the fungi *Candida, Yarrowia* and *Exobasidiomycetes. Tokophrya quadripartite* was higher abundant in activated sludge than in granular samples. Instead, *Vorticella aequilata* and *Trichodina sp.* were significantly abundant in the granules larger than 1.4 mm. This study shows for the first time the characterisation and comparison of the eukaryotic community in full-scale AGS systems.

In preparation: Mary Luz Barrios-Hernández, Hector Garcia, Berend Lolkema, Ben Abbas, Damir Brdjanovic, Mark C.M van Loosdrecht, Christine M Hooijmans

5.1 INTRODUCTION

Protozoa, fungi and metazoans are part of the large microbial eukaryotic community playing a significant role in wastewater treatment systems. They contribute to improve physicochemical and microbiological effluent quality parameter by enhancing the sludge production and sedimentation [108, 197, 198]; removing contaminants [199, 200], and significantly contributing to the pathogenic microorganisms removal [201, 202]. The evolutionary eukaryotic classification is continuously growing [134, 136, 203], helping to identify more extensive groups of organisms that are either present but not easy to perceive under (optical, fluorescent or scanning electron) microscopy [204, 205] or that they were morphologically classified into inaccurate taxonomic groups [206, 207].

In studying the presence of eukaryotes in wastewater treatment processes (anaerobic or aerobic), recent studies have been done using 18S rRNA gene sequencing analysis. The protist phyla Ciliophora and Cercozoan, and the fungi phyla Ascomycota and Basidiomycota were reported highly abundant in activated sludge systems [143, 145, 208]. For the AGS wastewater treatment process, limited information is available concerning eukaryotic assemblies. AGS is based on the agglomeration of functional bacteria agglomerated in the form of granules (> 0.2 mm) which simultaneously remove carbon and nutrients [13, 209]. A high abundance of stalked protozoa can be present on the surface of the granules, contributing to the removal of particulate solids and viral and bacterial organisms [93, 131].

Full-scale AGS systems (Nereda®) contain besides the larger granules, smaller bacterial agglomerations in the same biological tank. Ali, Wang [21] confirmed that the granular-sized agglomerations comprise different bacterial functional groups. Besides the eukaryotic organisms attached to the surface of the granules, it is thought that the free-living eukaryotes can find shelter between the smaller granules further contributing to the pathogen removal process (Chapter 3, Section 3.5). However, neither molecular composition, variations nor functionalities concerning the eukaryotic community in full-scale AGS systems have been yet studied.

The main objective of this chapter was to analyse the diversity, occurrence and variation of eukaryotic structures in the aerobic granules samples collected from two full-scale AGS systems using high-throughput 18S rRNA gene sequencing. Results were compared with parallel activated sludge systems treating the same raw wastewater influent as the AGS system. Moreover, an evaluation of the eukaryotic community classified into small or large granules was performed.

5.2 MATERIALS AND METHODS

5.2.1 Treatment facilities and sample collection

Samples (28) were collected over six months (December 2017–May 2018) from two full-scale WWTPs described in Chapter 2. Vroomshoop and Garmerwolde, the Netherlands. The raw sewage is divided over two parallel treatment systems which comprise of an AGS (Nereda®) process and an activated sludge system. Vroomshoop WWTP is a hybrid process, in which the excess sludge from the AGS system is discharged into a carrousel activated sludge configuration. Instead, Garmerwolde WWTP consists of two separate processes, an AGS and an adsorption/bio-oxidation (A&B) system. Configurations of the WWTPs can be seen in Figure 2.1. As described in Chapter 2, the treatment facilities showed stable operational conditions concerning COD, solids, nitrogen and phosphorous removal.

5.2.2 Sample collection and processing

Mixed liquor samples (1L) were taken from the raw sewage and the AGS and activated sludge reactors during the completely mixed aeration phase (Figure 5.1).

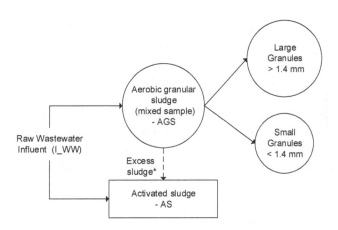

* Applicable only for Vroomshoop WWTP

Figure 5.1. Sample processing for raw influent wastewater, (mixed) granular and activated sludge samples taken from the combined WWTP facilities. (*) Excess sludge was applicable only for Vroomshoop WWTP.

Granular sludge samples were fractionated by sieving into small (< 1.4 mm) and large granules (> 1.4 mm). Additionally, the AGS matrices (mixed sample and large granules) were homogenised using a conventional kitchen blender. 2 mL samples were then washed twice in phosphate-buffered saline solution and centrifuged at 13.000 rpm for one minute to form pellets. Pellets were stored at -20 C approximately 2-3 years until performing the DNA extraction.

5.2.3 Microscopy and scanning electron observation

Granular and activated sludge samples were subjected to microscopy observations, including optical (X10 –X40) to check on the occurrence and activity of the organism. Scanning Electron Microscopy (SEM) was performed to catch and identify some organisms attached to the granular surfaces. SEM was carried out as described in Chapter 2.

5.2.4 DNA extraction and 18S rRNA gene sequencing

Genomic DNA was extracted from samples (0.2 - 0.3 g wet-weight) using QIAamp PowerFecal PRO DNA kit (QIAGEN) according to the manufacture's protocol. Samples were standardised to 1 ng/ μL using sterile water. The 18S rRNA gene sequencing was consistent with Barrios-Hernandez et al., *submitted*. In brief, the eukaryote-specific primers pair 528F 5'-GCGGTAATTCCAGCTCCAA-3' and 706R 5'-AATCCRAGAATTTCACCTCT-3' were used to amplify the V4 region of 18S rRNA genes using PCR. PCR products were purified for the library preparation. Paired-end reads and chimeras were removed/merged using FLASH (V1.2.7) and Qiime (Version 1.7.0). Uparse software (Uparse v7.0.1001) was used for the sequence analysis. Sequences with similarity ≥ 97% were assigned to the same Operational Taxonomic Units (OTUs). OTUs were compared with RDP Classifier (Version 2.2) and Silva database for species annotation with a threshold between 0.6 and 1 [210, 211]. The treated sequence data were uploaded to the National Center for Biotechnology (NCBI) under project accession number PRJNA671179.

5.2.5 Data analyses

Except for the raw wastewater influent (comprised only of two samples), groups were made of at least three (3) samples of either (mixed) granular sludge, large granules (> 1.4 mm), small granules (< 1.4 mm) or activated sludge. Alpha diversity indices (richness, diversity and sequencing depth) of the eukaryotic assemblages were processed using QIIME (Version 1.7.0) to estimate Shannon, Simpson, Chao1, Abundance-based Coverage Estimator (ACE) and Good's coverage indices. Wilcoxon rank-sum test was performed to determine significant differences among the richness and diversity of the groups. A p-value < 0.05 set the significance. OTUS were presented in Venn Diagrams as described in Shade and Handelsman [212].

A Beta-diversity analysis was carried out across raw wastewater and sludge groups (made of three samples) to determine whether the community structure significantly differed between groups. The analysis of molecular variance (AMOVA) was calculated by Mothur [213]. Bray-Adonis analysis was performed by R software (Version 2.15.3) using the Vegan package for the all samples (raw wastewater influent and sludge). For the sludge groups, Metastat analysis was implemented to distinguish the differences between the generic structures of the granular (mixed) samples and activated sludge at different taxa-levels. For Metastat the p-value was calculated by the method of permutation test while q-value was calculated by the method of Benjamini and Hochberg False Discovery Rate [214]. The significance was set at p-value < 0.05.

Additionally, the linear discriminant analysis (LDA) effect size (LEfSe) was performed to identify statistical differences among the abundant eukaryotic taxa. In this analysis, all group samples were included (mixed granules, small and large granules and activated sludge). An LDA score of 4 and p-value < 0.05 was considered to be significantly different [215].

5.3 RESULTS

In this study, raw wastewater influent, (mixed) granular sludge, large and small granules and activated sludge from the two WWTPs were analysed based on 18S rRNA gene sequencing. Prior to the analysis, sludge samples were observed under optical microscopy and SEM. Multiple eukaryotic structures were observed (free-swimming, and attached protozoa) and some of them could be made visible using SEM, especially the one attached to the granular surface. Figure 5.2 shows three different species of attached ciliates and how the granular sludge is used by the ciliate protozoa as attachment surface.

5.3.1 Diversity and composition per sample

A total of 3,143,210 sequences for the V4 region which corresponded to 1,691,460 for Vroomshoop and 1,451,750 for Garmerwolde WWTP were obtained (Annex, Table 5.3). The average species richness estimates (Shannon and Simpson) and the diversity analyses indexes (Chao1 and ACE) are shown in Table 5.1. Results indicated that the raw wastewater influent samples have the most extensive OTUs (larger diversity) when compared to the sludge samples. Apart from the Simpson index, the number of species observed in the influent raw wastewater were significantly higher (p-value $\ll 0.05$) than the rest of the granular and activated sludge samples at the two WWTPs (Vroomshoop and Garmerwolde). The full set of data per sample and p-values from the Wilcoxon rank-sum test are shown in Annex, Table 5.4 and Table 5.5).

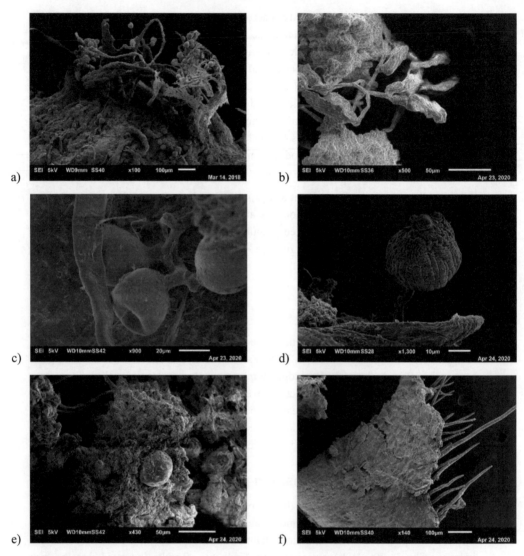

Figure 5.2 Eukaryotic structures detected using scanning electron microscopy in granules larger than 1.4 mm. a) Granular sludge surface, b), c) and d) attached ciliates from Oligohymenophorea class in the granular surface, e) non-identified finger-type structure.

Regarding the sludge samples, in Vroomshoop WWTP, the average Shannon, Chao1 and ACE indexes showed lower diversity in the (mixed) granular sludge (4.7, 485.9 and 521.7) than the activated sludge (4.9, 653.6 and 697.3). When comparing the granular fractions, the small granules < 1.4 mm (4.6, 632.7 and 671.3) showed more extensive estimations (higher diversity) than the granules larger than 1.4mm (4.6, 416.0 and 433.6). However, apart from the fraction with large granules (> 1.4 mm), which according to the Shannon-Wilcox index was significantly more diverse than activated sludge (p-value 0.0233), samples were equally distributed. Similar tendencies were observed for the Garmerwolde WWTP when comparing (mixed) granular sludge samples (3.7, 502.1 and 523.3) with activated sludge (4.7, 707.9 and 720.0). The fraction containing large granules seemed to be larger diverted (4.6, 651.2 and 652.0) than the fraction with small granules (4.2, 491.4 and 520.5), but all the sludge group samples were significantly different from each other.

Table 5.1 Species richness estimates and diversity indexes of the raw wastewater influent, granular and active sludge samples.

	Influent wastewater		raw	Mixed granular sludge			Large granules >1.4mm			Small granules < 1.4mm			Activated sludge		
Vroomshoop																
Shannon	5.5	±	0.4	4.7	±	0.4	4.6	±	0.2	4.6	±	0.3	4.9	±	0.2
Simpson	0.9	±	0.01	0.9	±	0.0	0.9	±	0.0	0.9	±	0.0	0.9	±	0.0
chao1	1080.1	±	148.8	485.9	±	166.3	416.0	±	145.6	632.7	±	41.3	653.6	±	48.3
ACE	1120.7	±	164.9	521.7	±	161.2	433.6	±	143.1	671.3	±	41.5	697.3	±	40.6
Garmerwolde																
Shannon	5.6	±	0.2	3.7	±	0.3	4.6	±	0.1	4.2	±	0.4	4.7	±	0.5
Simpson	0.9	±	0.0	0.8	±	0.0	0.9	±	0.0	0.9	±	0.0	0.9	±	0.1
chao1	1353.4	±	91.1	502.1	±	67.9	651.2	±	143.3	491.4	±	91.8	707.9	±	74.9
ACE	1390.7	±	86.9	523.3	±	70.7	652.0	±	142.8	520.5	±	86.1	720.0	±	75.3

5.3.2 Core microbiota

The unique and shared dominant OTUs of the four (4) grouped samples: raw wastewater influent, (mixed) granular sludge, large and small granules, and activated sludge are shown in Figure 5.3. In Vroomshoop WWTP, from the 1843 OTUs, 335 were found in at least one of the four samples. Granules (mixed samples) and activated sludge shared 544 OTUs.

The segregated samples that had more OTUs in common with the activated sludge were the fraction with small granules (527), followed by the fraction with larger granules (492).

In Garmerwolde WWTP, from 1873 OTUs, 365 were shared among all the samples. Granular sludge (mixed samples) shared 526 OTUs with activated sludge. The fraction with small granules shared 571 with activated sludge, while 567 with the larger-sized granular fraction. The most abundant phyla within this core group per WWTP is shown in Figure 5.4. The phylogenetic affiliation was categorised according to the phylum-level that belongs to either protist, fungi, metazoan, archaea or algae domains. Regardless of the location (Vroomshoop and Garmerwolde), organisms belonging to the unidentified Eukaryota group were the dominant protist community found in all samples followed by Ascomiycota (Fungi). Instead, Rotifera (Animalia) was more diverse, considering sludge samples instead of the influent.

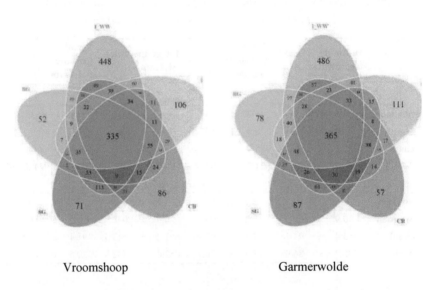

Vroomshoop   Garmerwolde

Figure 5.3. Venn diagrams of the unique and shared dominant OTUs between raw wastewater influent, I_WW (blue); (mixed) granular sludge (purple); small granules (darker purple); large granules (green); and activated sludge (orange).

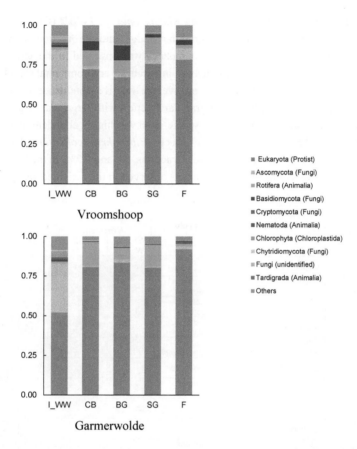

Figure 5.4 Stacked bar charts representing the more abundant 18S rRNA gene amplicons of the phyla-level (and domain) assigned in the raw wastewater influent (I_WW), (mixed) granular sludge (CB), large (BG) and small granules (SG) and activated sludge (F) samples of the target WWTPs.

5.3.3 Dominant eukaryotic structure community

In Figure 5.5, attention was given to species-level categorised into their genus and class levels of the more leading domain (> 1%) found with the 18S rRNA gene sequencing analysis. In Vroomshoop WWTP, from the Oligohymenophorea class, the dominant specie of the protist domain was *Opisthonecta henneguyi* with a relative abundance of 7.1% in the (mixed) granular sludge. It represented 2.2% in the fraction with larger granules (> 1.4 mm) and 1.2 % for the small granules (< 1.4 mm); while it was approximately 2.6% in the activated sludge.

Other abundant specie was the free living *Rhogostoma sp. 1966/2* (Carcozoa) reaching 2.1%, 1.3% and 2.6% in the (mixed) granular sludge, and the fractions with large and small granules, respectively. A higher abundance was observed in the activated sludge (3.1%). Regarding the specie identified as an invertebrate environmental sample from the Peronosporomycetes class, it reached an average relative abundance of 1.3% in the (mixed) granular sludge, 1.8% in the granular fraction with large granules and 2.4% in the fraction with smaller granules. This specie-level was highly abundant in the activated sludge (9.9%). Concerning fungi, the specie *Fibrophrys columna* (Basidiomycota phylum) was relatively high in the granular fractions, 5.5 and 9.2%, in the large and small granules, respectively. Lower abundance was observed in the activated sludge (2.74%).

Class (order)	Genus	Specie	Vroomshoop					Garmerwolde				
			IW W	C B	B G	S G	F	IW W	C B	B G	S G	F
Oligohyme nophorea	Colpidium	Colpidium striatum	0.6	0.0	0.0	0.1	0.0	1.6	0.1	0.1	0.1	0.0
	Opisthonecta	Opisthonecta henneguyi	0.2	7.1	2.2	1.7	2.6	0.2	0.7	0.5	0.5	0.5
		Vorticella microstoma	0.0	0.2	0.1	0.1	1.2	0.0	0.4	0.3	0.3	3.3
	Vorticella	Vorticella aequilata	0.3	0.5	1.5	1.1	0.4	0.5	2.8	3.4 8*	1.0	0.6
Peronos poromycetes	Unidentified Peronosporomycetes	Invertebrate environm.sampl	1.1	1.3	1.8	2.4	9.9	0.4	1.1	1.6	6.7	0.9
Phyllo pharyngea	Unidentified Conthreep	Trichodina sp.	0.0	0.7	1.0	0.3	0.3	0.1	0.2	2.5 9*	0.4	0.1
	Tokophrya	Tokophrya quadripartita	0.0	0.1	0.2	0.7	3.3	0.1	0.9	0.2	0.3	0.7
Arcellinida	Arcella	Arcella gibbosa	0.0	0.0	0.0	0.0	0.2	0.0	0.0	0.0	0.0	3.2
Unidentified Eukaryota	Gymnophrys	Athalamea environm.sampl	0.0	0.0	0.0	0.0	0.0	0.0	0.0	0.0	0.0	1.1
(Cryomonadida)	Rhogostoma	Rhogostoma sp.	9.0	2.1	1.3	2.6	3.1	1.0	0.9	3.4	14	18
(Saccharo -mycetales)	Candida	Candida austromarina	1.3	0.4	0.3	0.4	0.7	0.3	0.0	0.0	0.0	0.1
	Yarrowia	Yarrowia lipolytica	1.7	0.4	0.1	0.1	0.5	0.1	0.0	0.0	0.0	0.0
Exobasidio- mycetes	Unidentified Exobasidiomycetes	Fibrophrys columna	0.7	1.5	5.5	9.2	2.7	0.6	0.2	0.3	0.1	0.1

0%　　　　　　　　100%

Figure 5.5 Heat map of the relative abundance (%) assigned to the eukaryotic specie taxa-levels.

Raw wastewater influent (IWW), (mixed) granular sludge (CB), large granules (BG), small granules (SG), and activated sludge samples per WWTP. Significant differences are mark with an *

In Garmerwolde WWTP, the average relative abundance of the specie *Opisthonecta henneguyi* was lower than Vroomshoop with values lower than 1% for all the samples (mixed granular sludge, large and small granules and activated sludge). Instead, the stalked ciliate *Vorticella aequilata*, which was lower than 1% in all the Vroomshoop WWTP samples, showed higher relative abundance, being 2.8%, 3.5%, 1.0% and 0.6%. Regarding *Rhogostoma sp. 1966/2*, a lower relative abundance of 0.9% and 3.4% was determined for the (mixed) granular sludge and the fraction with larger granules, this when comparing with the smaller granules (14.2%) and activated sludge (17.9%). Concerning the invertebrate environmental sample specie, except for the fraction with smaller granules (6.7%), the rest of the samples showed a relative abundance lower than 1.6%. The fungi specie *Fibrophrys columna* was lower than 1% in all granular and activated sludge samples at Garmerwolde WWTP.

5.3.4 Variation among treatments

Significant differences for the eukaryotic community between the wastewater treatment processes were determined Overall, the Bray-Adonis analysis showed statistical differences ($p = 0.0014$) between the granular sludge groups (mixed, small and large granules) and activated sludge for both WWTPs. The comparison of the abundance between aerobic (mixed) granular sludge and activated sludge at different taxa-levels using Metastat analyses is shown in Table 5.2 for Vroomshoop and Garmerwolde, respectively.

In Vroomshoop WWTP, 11 genera and 12 species were different between treatment processes. Apart from the fungi specie *Pythiaceae sp*, the rest of the structures were significantly less abundant in AGS than activated sludge process. For Garmerwolde WWTP, Metastat showed differences in 23 genera and 14 species. The LEfSe analysis (Figure 5.6) confirmed that up to the specie-taxa level, in Vroomshoop WWTP, *Tokophrya quadripartite* (Phyllopharyngea class) was the specie-level making the most remarkable difference between the activated sludge and the rest of the granular samples. While in Garmerwolde, *Trichodina* sp. (Phyllopharyngea class) and *Vorticella aequilata* (Oligohymenophorea class) made the greatest difference in the fraction with larger granules (> 1.4 mm).

Table 5.2 Statistically different taxa abundance (median ± standard deviation, p-value< 0.05) between (mixed) granular sludge and activated sludge obtained from MetaStats analyses per WWTP.

Taxon	Combined granules			Activated sludge			p-value
Genera				Vroomshoop			
Prorodon	$9.06X10^{-04}$	±	$6.74X10^{-04}$	$9.68 X10^{-03}$	±	$2.11X10^{-03}$	0.0115
Arcella	$5.73X10^{-06}$	±	$5.73X10^{-06}$	$2.15X10^{-03}$	±	$4.65X10^{-04}$	0.0054
Tokophrya	$1.32X10^{-02}$	±	$4.52X10^{-03}$	$3.70X10^{-02}$	±	$7.03X10^{-03}$	0.0251
unidentified Triplonchida	0.00	±	0.00	$2.52 X10^{-04}$	±	$1.13X10^{-04}$	0.0498
Salpingoeca	$5.73X10^{-05}$	±	$5.73X10^{-05}$	$2.92X10^{-04}$	±	$1.72X10^{-05}$	0.0121
Heliophrya	0.00	±	0.00	$1.03 X10^{-04}$	±	$3.58X10^{-05}$	0.0232
Pythium	$9.23X10^{-04}$	±	$3.32X10^{-04}$	$2.08X10^{-03}$	±	$6.61X10^{-05}$	0.0174
Monoblepharis	$2.81X10^{-04}$	±	$6.61X10^{-05}$	$1.03X10^{-04}$	±	$2.63X10^{-05}$	0.0434
unidentified Pleosporales	$2.29X10^{-05}$	±	$1.52X10^{-05}$	$6.88X10^{-05}$	±	$9.93X10^{-06}$	0.0368
Cercomonas	$5.73X10^{-06}$	±	$5.73X10^{-06}$	$3.44X10^{-05}$	±	$9.93X10^{-06}$	0.0402
unidentified Xylariales	$5.73X10^{-06}$	±	$5.73X10^{-06}$	$2.87X10^{-05}$	±	$5.73X10^{-06}$	0.0264
Species							
Arcella gibbosa	$5.73X10^{-06}$	±	$5.73X10^{-06}$	$2.00X10^{-03}$	±	$4.20X10^{-04}$	0.0069
Tokophrya quadripartita	$7.51X10^{-04}$	±	$5.73X10^{-06}$	$3.30X10^{-02}$	±	$6.70X10^{-03}$	0.0062
Salpingoeca euryoecia	$4.59X10^{-05}$	±	$4.59X10^{-05}$	$2.35X10^{-04}$	±	$1.15X10^{-05}$	0.0089
Heliophrya erhardi	0.00	±	0.00	$1.03X10^{-04}$	±	$3.58X10^{-05}$	0.0226
Monoblepharis macrandra	$2.81X10^{-04}$	±	$6.61X10^{-05}$	$1.03X10^{-04}$	±	$2.63X10^{-05}$	0.0425
Pythium monospermum	$8.95X10^{-04}$	±	$3.38X10^{-04}$	$1.84X10^{-03}$	±	$1.24X10^{-04}$	0.0365
Salpingoeca ventriosa	$1.15X10^{-05}$	±	$1.15X10^{-05}$	$5.73X10^{-05}$	±	$1.52X10^{-05}$	0.0484
Knufia petricola	$2.29X10^{-05}$	±	$1.52X10^{-05}$	$7.45X10^{-05}$	±	$1.15X10^{-05}$	0.0307
Pleosporales sp. KF131	$1.72X10^{-05}$	±	$9.93X10^{-06}$	$4.59X10^{-05}$	±	$5.73X10^{-06}$	0.0411
Pythiaceae sp. PHY2	$2.29X10^{-05}$	±	$1.52X10^{-05}$	$9.17X10^{-05}$	±	$1.52 X10^{-05}$	0.0173
Seiridium sp. 51	$5.73X10^{-06}$	±	$5.73X10^{-06}$	$2.87X10^{-05}$	±	$5.73X10^{-06}$	0.0240
Arcella hemisphaerica	0.00	±	0.00	$1.49X10^{-04}$	±	$4.69X10^{-05}$	0.0180

Continuation Table 5.2 Statistically different taxa abundance (median ± standard deviation, p-value< 0.05) between (mixed) granular sludge and activated sludge obtained from MetaStats analyses per WWTP.

Taxon	Combined granules			Activated sludge			p-value
Genera				Garmerwolde			
Epistylis	$3.20X10^{-01}$	±	$7.13X10^{-02}$	$2.90X10^{-02}$	±	$1.27X10^{-02}$	0.0121
Vorticella	$2.85X10^{-02}$	±	$3.05X10^{-03}$	$6.21X10^{-03}$	±	$2.74 X10^{-03}$	0.0050
Trichosporon	$1.20X10^{-04}$	±	$9.93X10^{-06}$	$3.73X10^{-04}$	±	$5.82X10^{-05}$	0.0093
Fusarium	$2.81X10^{-04}$	±	$1.52X10^{-05}$	$6.48X10^{-04}$	±	$1.41X10^{-04}$	0.0336
Yarrowia	$5.73X10^{-06}$	±	$5.73X10^{-06}$	$2.52X10^{-04}$	±	$8.09X10^{-05}$	0.0213
Monosiga	$5.73X10^{-06}$	±	$5.73X10^{-06}$	$1.03X10^{-04}$	±	$2.98X10^{-05}$	0.0186
Monoblepharis	$6.25X10^{-04}$	±	$1.61X10^{-04}$	$9.75X10^{-05}$	±	$7.32X10^{-05}$	0.0235
unidentified_Streptophyta	0.00	±	0.00	$4.01X10^{-05}$	±	$1.15X10^{-05}$	0.0154
unidentified_Chlorophyceae	$4.01X10^{-05}$	±	$4.01X10^{-05}$	$2.12X10^{-04}$	±	$4.69X10^{-05}$	0.0284
Knufia	$1.55X10^{-04}$	±	$9.93X10^{-06}$	$2.12X10^{-04}$	±	$1.52X10^{-05}$	0.0197
Trithigmostoma	$1.43X10^{-04}$	±	$3.76X10^{-05}$	$1.15X10^{-05}$	±	$1.15X10^{-05}$	0.0175
Unident. Trebouxiophyceae	$4.59X10^{-05}$	±	$1.15X10^{-05}$	$2.18X10^{-04}$	±	$7.59X10^{-05}$	0.0480
Acineria	$6.31X10^{-05}$	±	$3.19X10^{-05}$	$5.05X10^{-04}$	±	$1.90X10^{-04}$	0.0464
Parastagonospora	$6.31X10^{-05}$	±	$3.03X10^{-05}$	$1.78X10^{-04}$	±	$2.50X10^{-05}$	0.0251
Boeremia	$5.73X10^{-05}$	±	$1.52X10^{-05}$	$1.15X10^{-04}$	±	$2.07X10^{-05}$	0.0485
SCM16C66	$8.03X10^{-05}$	±	$2.07X10^{-05}$	$2.29X10^{-05}$	±	$5.73X10^{-06}$	0.0306
Valsa	$1.72X10^{-05}$	±	$9.93X10^{-06}$	$4.59X10^{-05}$	±	$5.73X10^{-06}$	0.0397
Cryptocaryon	$1.38X10^{-04}$	±	$3.97X10^{-05}$	$5.28X10^{-04}$	±	$1.06X10^{-04}$	0.0170
Pilobolus	0.00	±	0.00	2.87X10-05	±	1.15X10-05	0.0352
Scopulariopsis	0.00	±	0.00	1.72X10-05	±	0.00E+00	0.0000
Picochlorum	0.00	±	0.00	4.59X10-05	±	1.52X10-05	0.0230
Myrmecia	0.00	±	0.00	1.72X10-05	±	0.00	0.0000
unidentified_Ulvophyceae	0.00	±	0.00	2.29X10-05	±	5.73X10-06	0.0126

Continuation Table 5.2 Statistically different taxa abundance (median ± standard deviation, p-value< 0.05) between (mixed) granular sludge and activated sludge obtained from MetaStats analyses per WWTP.

Taxon	Combined granules			Activated sludge			p-value
species				Garmerwolde			
Vorticella aequilata	2.82×10^{-02}	±	2.95×10^{-03}	5.87×10^{-03}	±	2.64×10^{-03}	0.0060
Yarrowia lipolytica	5.73×10^{-06}	±	5.73×10^{-06}	2.52×10^{-04}	±	8.09×10^{-05}	0.0219
Monosiga ovata	5.73×10^{-06}	±	5.73×10^{-06}	1.03×10^{-04}	±	2.98×10^{-05}	0.0200
Monoblepharis macrandra	6.25×10^{-04}	±	1.61×10^{-04}	9.75×10^{-05}	±	7.32×10^{-05}	0.0250
Vermamoeba vermiformis	4.01×10^{-05}	±	1.52×10^{-05}	5.73×10^{-06}	±	5.73×10^{-06}	0.0461
Trithigmostoma steini	1.43×10^{-04}	±	3.76×10^{-05}	1.15×10^{-05}	±	1.15×10^{-05}	0.0194
Valsa mali	1.72×10^{-05}	±	9.93×10^{-06}	4.59×10^{-05}	±	5.73×10^{-06}	0.0348
Pilobolus sphaerosporus	0.00	±	0.00	2.87×10^{-05}	±	1.15×10^{-05}	0.0299
Scopulariopsis brevicaulis	0.00	±	0.00	1.72×10^{-05}	±	0.00	0.0000
Picochlorum eukaryotum	0.00	±	0.00	4.59×10^{-05}	±	1.52×10^{-05}	0.0238
Colpodellidae sp. HEP	4.01×10^{-05}	±	1.15×10^{-05}	1.15×10^{-05}	±	5.73×10^{-06}	0.0429
Myrmecia astigmatica	0.00	±	0.00	1.72×10^{-05}	±	0.00	0.0000
Pirula salina	0.00	±	0.00	2.29×10^{-05}		5.73×10^{-06}	0.0121
Scenedesmus sp. KMMCC 1534	0.00	±	0.00	2.29×10^{-05}	±	5.73×10^{-06}	0.0121

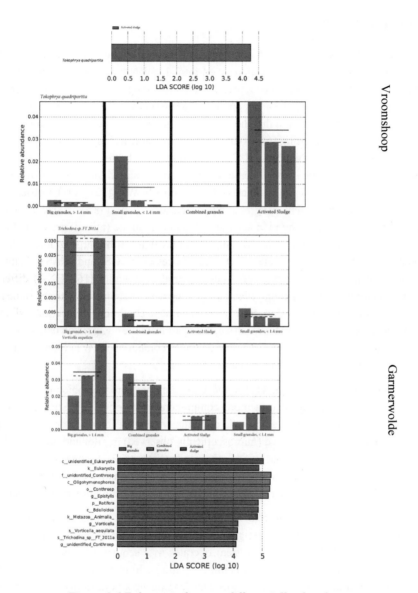

Figure 5.6 Eukaryotic features differentially abundance.

(Mixed) granular sludge, large granules (>1.4 mm) and activated sludge at Vroomshoop WWTP. Calculation was based on the linear discriminant analysis (LDA) effect size.

5.4 DISCUSSION

To elucidate the composition and dominance of the eukaryotic structures in the AGS treatment systems, genetic structures derived from 18S RNA gene sequencing analyses were determined. A comparison with parallel activated sludge treatments was performed. Regardless the location (Vroomshoop and Garmerwolde, the Netherlands), results presented in Figure 5.3 showed that the raw wastewater influent has the most diverse and unique OTUs, vastly diverting ($p \ll 0.05$) from the biological sludge samples [163]. Our results were opposite to the previous findings reported for bacterial communities [216], in which activated sludge showed higher richness and evenness (diversity) than the raw wastewater influent. In our study, in accordance with Ali, Wang [21], OTUs observed in the raw wastewater influent progressively decreased as the granular size increased. Unique OTUS were more found in activated sludge than in the (mixed) granular sludge, and they were more abundant in the small granules than in the large granules. When comparing richness and evenness between processes (AGS and CAS), Winkler, Kleerebezem [217] revealed that the AGS bacterial community (16S DNA) differ in distribution from activated sludge, even though they were systems fed with the same raw wastewater influent (composition, flow rate and climatic conditions). Our results showed that the eukaryotic diversity indexes varied, but apart from the comparison between larger granules and activated sludge ($p = 0.023$) in Vroomhooop WWTP, no significant differences were found between the sludge samples.

Based on the shared and unique OTUs, the Venn diagram core microbiome represented structures that were assigned to a more comprehensive taxon [212]. Such taxonomic classification showed a larger eukaryotic community (protist, fungi, chloroplastida, and animalia) than either the reported (microscopically observed) in pilot and full-scale AGS systems [75, 85]; or the ones observed in this study (Figure 5.2). Some species are well-described in the literature regarding morphology and functionality [129, 218, 219], but some are difficult to distinguish microscopically because of their similar appearances.

Overall, the 18S gene eukaryotic assemblies (specifically protist and fungi) here presented followed previous investigations in wastewater treatments using 18S gene sequencing [145, 220, 221]. AGS system operates as a sequence batch reactor (SBR) while activated sludge operates in a continuous mode. In our study, regardless of the system type or location, protists were indeed the leading eukaryotic domain in all samples (Figure 5.4). Among them, ciliates were found (stalked, free-swimming and crawling protozoa) from the class Oligohymenophorea (Figure 5.5) which are indicators of a good treatment performance [141, 142, 222]. The bacterivorous *Vorticella aequilata* (*Vorticella*) form a contractile stalk to bind surfaces [223]. It showed to be significantly dominant (Figure 5.6) in the granular samples (mixed and granules > 1.4 mm) in Garmerwolde WWTP [224, 225].

For *Vorticella microstoma* (*Opisthonecta*), it was expected to observe similar behaviour in the granular surfaces as in the activated sludge. *V. microstoma* is a very resistant protist recognised for taking up heavy metals such as lead, cadmium, copper, nickel and mercury ions [226, 227]. It also forms stalks after their free-swimming stage, however, Bramucci and Nagarajan [228] showed that in presence of inhibitors they could stay in such stage. Interestingly, activated sludge was more favourable for *V. microstoma* at both WWTPs (3.2% in Garmerwolde and 1.2% in Vroomshoop) than granular sludge (< 1% in all the samples). Likewise, the closely related *Opisthonecta henneguyi,* which is a permanent motile (ciliate) member of the peritrich subclass [206, 229], was particularly abundant in granular sludge (7.1% in the mixed sample) when compared with activated sludge (2.6% at Vroomshoop WWTP), and relatively low (< 1%) in all samples from Garmerwolde WWTP. Variations of typical eukaryotic structures and their behaviour (occurrence and abundance) could be related to factors such as diet, environmental conditions and oxygen preferences [126, 218, 230]. From the operational point of view, in the AGS reactor, anaerobic and aerobic zones are fixed per operational cycle, and high shear forces are present. In an activated sludge system different oxygen zones are occurring as function of time and location in the reactor. Vroomshoop WWTP has the particularity of discharging excess granular sludge to the activated sludge process (carrousel configuration), therefore, to some extend migration of the structures from AGS to the activated sludge systems was expected. Little is known about the effect of operational conditions on the development of eukaryotes in wastewater treatment systems; hence, a different research approach is recommended to determine the effect of operational parameters in their occurrence.

According to the Sludge Biotic Index proposed by Madoni [222], a high abundance of small flagellates and free-swimming ciliates in activated sludge systems can be indicators of moderate treatment performance. However, even though small species were abundant in both WWTPs, a good performance was always recorded during this study [98]. One if the crawling and free-living ciliates, is the facultative ectosymbiotic suctorian *Tokophrya quadripartite* (Phyllopharyngea class). It marked a difference between granular sludge and activated sludge in Vroomshoop WWTP, which in agreement to Madoni and Ghetti [231], lower abundance can be found in biofilm-type systems (rotating disk) than in activated sludge. Its abundance can also be related to seasonality. For example, carnivorous *Tokophrya* genus (fed on motile ciliates) has been mostly observed during periods of low temperatures [232]. In contrast, the crawling parasitic *Trichodina sp. FT-2011a*, which was more abundant in the granules > 1.4 mm than in the rest of the samples in Garmerwolde WWTP, can succeed better during warmer periods [233]. Our study was carried out in samples taken during autumn and winter (10-15 °C), therefore, changes in the eukaryotic structure's abundance could be expected during summer.

With respect to other free-living organisms, *Rhogostoma sp.* was a remarkable abundant genus in the small granular (<1.4 mm) fraction and activated sludge of both WWTPs. This genus is common in WWTPs [145, 163] but the reason behind its high abundance is still unknown. Öztoprak, Walden [138] associated some species from the Rhogostomidae community with biomass production in the WWTPs. Attention is mostly given to their occurrence in soils, root and leaves [140]. Further studies are required to better determine their role and other protists' functionalities in AGS systems.

Concerning fungi, previous identification has been carried out in laboratory-scale and full-scale AGS systems [234-236]. As in this study, the phylum Ascomycota has been reported as the dominant group in AGS and in other wastewater treatment systems [237]. One of the Ascomycota, the genus *Candida,* is a recognised contributor of contaminants removal such as chromium, hydrocarbons, phenols, but also related with bulking sludge, biofilm production and foaming in WWTP [199, 200, 238]. From the second most abundant phyla Basidiomycota, the specie *Fibrophrys columna* was highly abundant in Vroomshoop WWTP. This specie was cultivated from fresh water and sequenced for the first time by Takahashi, Yoshida [239]. However, species from the *Amphifilidae* group remain still unknown; to the author's knowledge, no other literature about its occurrence in wastewater is available.

5.5 CONCLUSIONS

In this study, the diversity and abundance of eukaryotic structures in aerobic granular sludge were compared with the activated sludge system fed with the same municipal wastewater raw influent. This study provides information on the eukaryotic composition and dominant (genera and species) taxa-level contained in full-scale AGS systems and activated sludge systems. Multiple species, in special from the protist group, were statically different among treatments (granular and activated sludge) but not with an abundancy higher than 1 %. To our knowledge, it is the first time the characterisation of the eukaryotic community associated with full-scale aerobic granular sludge is provided. However, the role and contribution of eukaryotic structures in the effluent purification may require further studies.

5.6 ANNEX

Table 5.3 Raw sequencing read and after QC sequencing reads of the two target treatment plants.

	Sample ID	Raw PE(#)	Qualified(#)	Nochime(#)	AvgLen(nt)	Q20	Q30	Effective%
Vroomshoop	M33.I	135,131	131,041	127,385	305	98.88	96.38	94.27
	M34.I	97,119	94,374	87,907	301	98.97	96.64	90.51
	M04.BG	128,186	125,276	122,384	297	99.05	96.92	95.47
	M08.BG	132,661	128,774	125,970	306	98.76	96.02	94.96
	M14.BG	136,745	133,772	130,813	301	99.03	96.79	95.66
	M05.SG	98,542	95,333	92,906	306	98.96	96.58	94.28
	M07.SG	138,809	134,652	131,118	298	98.96	96.68	94.46
	M10.SG	130,837	126,835	123,143	307	98.77	96.05	94.12
	M09.CB	149,004	144,843	141,141	300	98.95	96.6	94.72
	M12.CB	123,062	117,855	114,095	303	98.89	96.35	92.71
	M13.CB	110,814	108,320	101,921	298	99.01	96.78	91.97
	M01.F	142,238	138,701	135,710	303	98.96	96.64	95.41
	M06.F	126,162	121,505	119,218	307	98.90	96.41	94.50
	M11.F	147,567	142,060	137,749	301	98.99	96.64	93.35
	Total	**1,796,877**	**1,743,341**	**1,691,460**				
Garmerwolde	M31.I	129,350	125,428	117,134	299	98.97	96.7	90.56
	M32.I	132,050	128,801	121,559	302	99.01	96.74	92.06
	M20.BG	134,031	130,575	126,029	294	98.90	96.42	94.03
	M22.BG	135,018	123,541	115,225	300	98.72	95.88	85.34
	M24.BG	84,430	82,402	79,991	290	98.91	96.44	94.74
	M21.SG	134,403	129,327	126,734	309	98.48	95.2	94.29
	M23.SG	135,824	127,771	124,718	311	98.65	95.57	91.82
	M28.SG	66,446	64,443	63,144	302	98.82	96.24	95.03
	M16.CB	92,792	90,527	88,046	290	98.98	96.67	94.89
	M17.CB	78,521	74,973	72,394	297	98.93	96.47	92.20
	M19.CB	149,029	144,507	139,811	291	98.89	96.44	93.81
	M15.F	104,973	102,448	100,155	299	98.93	96.49	95.41
	M18.F	94,416	91,339	88,712	301	98.69	95.85	93.96
	M27.F	92,386	89,378	88,098	301	98.78	96.22	95.36
	Total	**1,563,669**	**1,505,460**	**1,451,750**				

Table 5.4 Alpha diversity index of the influent raw wastewater, aerobic granules, big and small granules and activated sludge samples from Vroomshoop wastewater treatment plants.

	Sample ID	Sample	Observed species	Shannon	Simpson	chao1	ACE	Goods coverage	PD whole_tree
	M33.I	Influent	783	5.033	0.916	931.298	955.826	0.996	91.897
	M34.I	Influent	1138	5.905	0.926	1228.837	1285.627	0.996	131.46
	M04.BG	Big granules	357	4.563	0.923	415.952	433.617	0.999	47.155
	M08.BG	Big granules	529	4.602	0.913	712.068	726.146	0.997	63.908
	M14.BG	Big granules	354	4.113	0.881	392.042	412.754	0.999	54.581
Vroomshoop	M05.SG	Small granule	460	4.581	0.925	589.493	602.982	0.998	57.441
	M07.SG	Small granule	523	4.160	0.876	632.698	671.302	0.997	65.857
	M10.SG	Small granule	545	4.847	0.930	690.283	702.200	0.997	64.778
	M09.CB	Combined granules	417	4.680	0.923	485.908	521.696	0.998	47.167
	M12.CB	Combined granules	645	5.394	0.932	818.298	827.329	0.997	82.036
	M13.CB	Combined granules	365	4.528	0.922	448.143	457.871	0.998	54.904
	M01.F	Activated sludge	469	4.828	0.916	567.063	622.603	0.998	59.193
	M06.F	Activated sludge	559	4.870	0.918	680.143	716.829	0.997	69.784
	M11.F	Activated sludge	518	5.238	0.934	653.625	697.278	0.997	69.063

Table 5.5 Alpha diversity index of the influent raw wastewater, aerobic granules, big and small granules and activated sludge samples from Garmerwolde wastewater treatment plants.

	Sample ID	Sample	Observed species	Shannon	Simpson	chao1	ACE	Goods coverage	PD whole_tree
	M31.I	Influent	1205	5.999	0.932	1535.587	1564.539	0.994	145.182
	M32.I	Influent	1003	5.135	0.884	1171.286	1216.833	0.996	120.437
	M20.BG	Big granules	500	4.601	0.918	651.185	652.049	0.997	71.714
	M22.BG	Big granules	592	4.686	0.920	844.897	865.551	0.996	74.649
	M24.BG	Big granules	448	4.531	0.912	494.511	518.841	0.998	70.451
	M21.SG	Small granule	395	4.196	0.869	491.397	501.365	0.998	53.852
Garmerwolde	M23.SG	Small granule	403	4.094	0.845	485.975	520.476	0.998	55.376
	M28.SG	Small granule	615	4.909	0.926	683.343	692.892	0.998	79.903
	M16.CB	Combined granules	401	4.423	0.911	502.061	523.315	0.998	65.707
	M17.CB	Combined granules	429	3.702	0.815	487.367	513.005	0.998	64.043
	M19.CB	Combined granules	486	3.706	0.802	638.167	667.938	0.997	60.575
	M15.F	Activated sludge	463	4.686	0.921	635.667	649.689	0.997	68.864
	M18.F	Activated sludge	577	4.777	0.914	707.900	720.045	0.997	79.983
	M27.F	Activated sludge	598	3.640	0.810	817.804	832.575	0.996	73.255

6

OUTLOOK AND CONCLUSIONS

The research described in this thesis focuses on determining the mechanisms involved in the removal and degradation of enteric viral and bacterial indicator organisms in AGS systems. This section provides a critical reflection on the results obtained in this investigation, their limitations and challenges. Opportunities for future research are also discussed.

6.1 INSIGHTS INTO THE PATHOGEN DEGRADATION PROCESS

6.1.1 The fate of the pathogens and faecal indicators after treatment

In this thesis, the removal efficiencies of faecal indicator organisms were determined in two combined wastewater treatment plants comprising both an AGS system and a CAS process. Findings in this dissertation showed that the AGS technology, operated as a sequencing batch reactor, can achieve similar removal efficiencies of faecal indicators as the more complex CAS systems; this was for the studied carrousel configuration as well as for the adsorption/bio-oxidation treatment process (Chapter 2). Using faecal indicators was sufficient to answer our first research question of comparing the treatment plants' processes on the faecal contaminant's removals [52, 61].

The faecal indicators concentrations measured in the treated effluent confirmed that they are not entirely removed from the raw wastewater, regardless of the type of treatment. As a result, adverse consequences for public health can be expected by the pathogens, virulent genes and antibiotic-resistant bacteria present in the discharged water.

However, literature has shown that faecal indicators do not always straightforwardly correlate with actual pathogens [58, 240]. The number of pathogenic organisms can be significantly higher depending on local infections, and even the detection methods used can provide different outcomes. Furthermore, attention should be given to pathogens that can grow in wastewater treatment plants, such as *Legionella* spp. It is different from typical waterborne pathogens as the exposure route is via aerosols. *Legionella* is a bacterium highly related to environmental water contamination [241]. Caicedo, Rosenwinkel [242] associates municipal wastewater treatment plants (WWTP) with the persistence of *Legionella* due to the optimal growth environment (i.e., temperature > 30◦C, high nitrogen concentrations, and potential host-related organisms).

Also, critical priority pathogenic organisms responsible for high morbidity and mortality rates such as carbapenem resistance bacteria (*Acinetobacter baumannii* and *Pseudomonas aeruginosa*) *Enterobacteriaceae* and their virulence genes require attention [243].

Besides measuring pathogen removal, measuring specific pathogens in the influent can reveal relevant information on the health and infection rate of a population, as is the case with the non-water-borne SARS-cov-2. This virus is nowadays highly abundant in sewage, and is used to monitor the infection rate of the population [244, 245].

6.1.2 Stable granulation with an overgrowth of protozoa in the surfaces

Experiments carried out in laboratory-scale systems fed with a mixture of (synthetic) wastewater and faecal sludge (Chapter 4) led to a better understanding of the formation and morphology of the granules at 20∘C; hence, an association between the growth of stalked ciliates and solids in the influent was established. Besides analysing the reactor performance, the use of synthetic faecal sludge aimed at supporting the requirements of countries where onsite sanitation systems are provided.

The addition of particulate material in the reactors promoted the overgrowth of protozoa, contributing to removing solids. It was expected that the high abundance of protozoa would also result in the removal of a considerable number of pathogens. However, this was not measured when such excessive abundance of stalked ciliates occurred. More research is required on the prey-predator relationships. Pure cultures of selected organisms from the protist domain may help to clarify their role in different AGS systems and identify their relationships with other (indicator/pathogenic) organisms.

Moreover, besides protists, other eukaryotes are reported for AGS systems. Sharaf, Guo [235] described an overgrowth of the filamentous *Geotrichum* (an acid-tolerant yeast-like fungus) in a laboratory-scale reactor fed with acetate. The authors associated the filamentous overgrowth with granular disintegration along with a disturbance in the bacterial community, which impacted the phosphorus removal rates. Previous literature suggested that particulate material (yeast) would induce finger-type organisms to grow [11] making the granular system unstable. In our study, we added the particulate material after granules were formed on acetate, and the system remained stable. The granules got reduced in their size and were fully covered by stalked ciliates, but no fluffy granules were observed. A good chemical performance, mainly based on chemical oxygen demand and phosphorus removal, guaranteed the system stability. Long operational times were required to achieve nitrification/denitrification.

In our study, using a faecal sludge simulant gave insight in the behaviour of the AGS system regarding granular stability and the effect on the eukaryotic community. It is possible that the bacterial and eukaryotic community changed to comply with the new influent characteristics and reactor operational conditions. High throughput screening will help in the assessment and the comparison with samples from real-life applications. Moreover, further studies such as the combined effect of faecal sludge addition and increased temperature (30∘C) on the eukaryotic community and pathogenic/surrogates organisms' removal are needed.

6.1.3 Pathogens and faecal indicators removal mechanisms

Pathogens and faecal indicators their association with particles (solids)

Based on the results from full-scale and laboratory-scale AGS systems, physical and biological mechanisms related to the removal dynamics were determined (Chapter 3). Granules formed only with synthetic wastewater showed a lower abundance of predators. Their low abundance made it possible to evaluate mechanisms as the attachment of faecal surrogates (*E. coli* and MS2 bacteriophages) to the granules. Results showed that the surrogates could saturate the granular surface, protecting themselves for being predated, being a significant outcome with negative implications for the public health in case of reuse of sludge. For the effluent quality, however, it is a positive outcome. Studying the effect of different granules (size, composition) could help to better understand the granules' adsorption capacity and how it impacts pathogen removal. Predation by higher-level organisms and cell lysis might disturb the investigation of the adsorption process. This can be avoided using eukaryotic inhibitors and controlled environments to avoid viral replication.

In order to enhance pathogen removal by adsorption, studying the effect that organic and inorganic compounds can have on the absorption capacity of the granules might be interesting. According to Wang, Liu [246], compounds usually found in domestic wastewater such as $NaCl$, $CaCl_2$ and H_2SO_4, can chemically modify the surface charges, increasing the pathogen attachment capacity. Co-treating saline water or high organic and inorganic compounds could be an option for full-scale municipal systems.

Finally, further study is needed to determine the impact of solids when post-treatment for effluent reuse is required. The AGS process shows higher solid concentrations in the effluent than CAS [96], which might hinder post-treatment or ultraviolet (UV) radiation [72, 247].

Removal by predation

Predation by protozoa is more efficient than adsorption for the removal of pathogens from the liquid bulk during the aeration phase, which was confirmed in the AGS laboratory-scale reactors (Chapter 3). The fact that overgrowth of protozoa did not occur when systems were fed with synthetic influent, either spiked with E. coli bacteria or MS2 bacteriophages, led to having a better understanding of their food selection pressure. Under the microscope stalked ciliates were visible. Bacterivorous genera Epistylis, Pseudovorticella, Vorticella and Vorticellides from Oligohymenophorea class took advantage of the granular surface and contributed to the removal of suspended bacteria.

The advanced high throughput screening also helped to identify the eukaryotes that could not be detected by optical microscopy. The study revealed that free-living organisms such as the genera Rhogostoma (Cercozoa) and Telotrochidium (Oligohymenophorea) participated in the removal process. Moreover, it was assumed that smaller particles such as viruses are normally taken as "bycatch" in the system. The removal of MS2 bacteriophages was significantly lower than the E. coli removal, and was related to a diverse but lower protozoa abundance. Concerning public health, those findings have broader applicability to explain why viruses are so persistent in wastewater and the difficulties to remove them from treated effluents. [120]. Further treatment is required for their elimination, justifying the need for post-treatment in case of effluent reuse, as previously mentioned in Section 6.1.1.

6.1.4 Protozoa predators, who are they?

Nowadays, the number of studies on the protozoa and fungi in relation to the granular formation and their association with the viral and bacterial community is increasing [234, 235]. Protozoans (of the protist group) were expressively significant in our study, and mostly responsible for the faecal indicator removal in full and laboratory-scale systems. Moreover, they contributed to the removal of particulate material.

A comprehensive investigation to understand their occurrence and diversity was carried out in two combined full-scale biological processes (AGS and CAS), identifying the eukaryotes in the different sludge matrices (Chapter 5). That included a mixture of granular sludge, samples with either large or small granular fraction, and activated sludge. Larger eukaryotic categories from class to species-level and diverted structures were found when comparing with previously studies in laboratory-scale reactors (Chapter 3). The eukaryotic composition, entering the systems via the raw wastewater, shifted according to the system demands. Like for bacterial communities, it was assumed that they adapted to their new ecosystems, being surrounded by new competitors and a different media composition [21].

However, when comparing activated and granular sludge, no strong differences were found. For example, the attached ciliated *Vorticella aequilata* and the crawling parasitic *Trichodina* sp. were significantly more dominant in the AGS wastewater treatment system than in the CAS process. But that was measured only for the larger treatment plant in which both systems (CAS and AGS) are operated completely separated. The above-mentioned organisms did not show differences in the smaller WWTP, but for this one the surplus sludge of the AGS reactor is discharged in the CAS system.

Overall, identifying the eukaryotes in AGS systems and comparing them with CAS sytems gives additional insight in the granular composition in full-scale applications. Investigations in laboratory-scale reactors have shown that predators can modulate the bacterial composition and affect granular formation [94, 248]. Futher studies on their contribution are relevant, like a study on hosting leading eukaryotes as free-living amoebae for *Legionella* bacteria to help controlling local infections and their virulence genes' promotion [249-251].

REFERENCES

1. van Loosdrecht, M.C. and D. Brdjanovic, *Anticipating the next century of wastewater treatment.* Science, 2014. **344**(6191): p. 1452-1453.

2. Sabliy, L., et al., *New approaches in biological wastewater treatment aimed at removal of organic matter and nutrients.* Ecological Chemistry and Engineering S, 2019. **26**(2): p. 331-343.

3. Orhon, D., *Evolution of the activated sludge process: the first 50 years.* Journal of Chemical Technology and Biotechnology, 2015. **90**(4): p. 608-640.

4. Van Haandel, A. and J. Van Der Lubbe, *Handbook biological waste water treatment-design and optimisation of activated sludge systems.* 2007: Webshop Wastewater Handbook.

5. Henze, M., et al., *Biological wastewater treatment: principles, modelling and design.* 2008: IWA publishing.

6. Vaiopoulou, E. and A. Aivasidis, *A modified UCT method for biological nutrient removal: Configuration and performance.* Chemosphere, 2008. **72**: p. 1062-8.

7. Dold, P., C. Bye, and Z.-r. Hu, *Nutrient Removal MBR Systems: Factors in Design and Operation.* Proceedings of the Water Environment Federation, 2010. **2010**: p. 5839-5852.

8. Heijnen, J. and M. Van Loosdrecht, *Method for acquiring grain-shaped growth of a microorganism in a reactor.* Biofutur, 1998. **183**(1998): p. 50.

9. de Kreuk, M.K., N. Kishida, and M.C.M. van Loosdrecht, *Aerobic granular sludge – state of the art.* Water Science and Technology, 2007. **55**(8-9): p. 75.

10. Morgenroth, E., et al., *Aerobic granular sludge in a sequencing batch reactor.* 1997. **31**(12): p. 3191-3194.

11. Beun, J., et al., *Aerobic granulation in a sequencing batch reactor.* Water Research, 1999. **33**(10): p. 2283-2290.

12. De Bruin, L., et al., *Aerobic granular sludge technology: an alternative to activated sludge?* Water Science & Technology, 2004. **49**(11): p. 1-7.

13. Sepúlveda-Mardones, M., et al., *Moving forward in the use of aerobic granular sludge for municipal wastewater treatment: an overview.* Reviews in Environmental Science and Bio/Technology, 2019. **18**(4): p. 741-769.

14. Dangcong, P., et al., *Aerobic granular sludge—a case report.* Water Research, 1999. **33**(3): p. 890-893.

15. Khan, M.Z., P.K. Mondal, and S. Sabir, *Aerobic granulation for wastewater bioremediation: a review.* The Canadian Journal of Chemical Engineering, 2013. **91**(6): p. 1045-1058.

16. Nancharaiah, Y.V. and G. Kiran Kumar Reddy, *Aerobic granular sludge technology: Mechanisms of granulation and biotechnological applications.* Bioresource Technology, 2018. **247**: p. 1128-1143.

17. de Sousa Rollemberg, S.L., et al., *Aerobic granular sludge: Cultivation parameters and removal mechanisms.* Bioresource Technology, 2018. **270**: p. 678-688.

18. Bueno, R.d.F., et al., *Simultaneous removal of organic matter and nitrogen compounds from landfill leachate by aerobic granular sludge.* Environmental Technology, 2020: p. 1-15.

19. Kishida, N., et al., *Anaerobic/oxic/anoxic granular sludge process as an effective nutrient removal process utilizing denitrifying polyphosphate-accumulating organisms.* Water Research, 2006. **40**(12): p. 2303-2310.

20. Lee, B.J., et al., *Measurement of Ordinary Heterotrophic Organism Active Biomass in Activated Sludge Mixed Liquor: Evaluation and Comparison of the Quantifying Techniques.* Environmental Engineering Research, 2014. **19**(1): p. 91-99.

21. Ali, M., et al., *Importance of species sorting and immigration on the bacterial assembly of different-sized aggregates in a full-scale aerobic granular sludge plant.* Environmental science & technology, 2019. **53**(14): p. 8291-8301.

22. Pronk, M., et al., *Effect and behaviour of different substrates in relation to the formation of aerobic granular sludge.* Applied Microbiology and Biotechnology, 2015. **99**(12): p. 5257-5268.

23. Liu, Y. and J.-H. Tay, *State of the art of biogranulation technology for wastewater treatment.* Biotechnology Advances, 2004. **22**(7): p. 533-563.

24. Corsino, S.F., et al., *Effect of extended famine conditions on aerobic granular sludge stability in the treatment of brewery wastewater.* Bioresource Technology, 2017. **226**: p. 150-157.

25. de Kreuk, M.K., M. Pronk, and M.C. van Loosdrecht, *Formation of aerobic granules and conversion processes in an aerobic granular sludge reactor at moderate and low temperatures.* Water Research, 2005. **39**(18): p. 4476-84.

26. Di Bella, G. and M. Torregrossa, *Simultaneous nitrogen and organic carbon removal in aerobic granular sludge reactors operated with high dissolved oxygen concentration.* Bioresource Technology, 2013. **142**: p. 706-713.

27. Bassin, J.P., et al., *Measuring biomass specific ammonium, nitrite and phosphate uptake rates in aerobic granular sludge.* Chemosphere, 2012b. **89**(10): p. 1161-1168.

28. de Kreuk, M.K., J.J. Heijnen, and M.C.M. van Loosdrecht, *Simultaneous COD, nitrogen, and phosphate removal by aerobic granular sludge.* Biotechnology and Bioengineering, 2005a. **90**(6): p. 761-769.

29. Liu, W., F. Yin, and D. Yang, *Granules abrasion cause deterioration of nitritation in a mainstream granular sludge reactor with high loading rate.* Chemosphere, 2020. **243**: p. 125433.

30. de Kreuk, M.K., J.J. Heijnen, and M.C.M. van Loosdrecht, *Simultaneous COD, nitrogen, and phosphate removal by aerobic granular sludge.* Biotechnology and Bioengineering, 2005. **90**(6): p. 761-769.

31. Schmeller, D.S., F. Courchamp, and G. Killeen, *Biodiversity loss, emerging pathogens and human health risks.* 2020, Springer.

32. Gao, G.F., *From "A" IV to "Z" IKV: attacks from emerging and re-emerging pathogens.* Cell, 2018. **172**(6): p. 1157-1159.

33. Kirhensteine, I., et al., *EU-level instruments on water reuse. Final report to support the commission's impact assessment.* 2016, European Union: Luxemburgo.

34. CDC. *Waterborne Disease in the United States.* 2020 December 1, 2020 [cited 2021.

35. Lodder, W. and A. de Roda Husman, *Presence of noroviruses and other enteric viruses in sewage and surface waters in The Netherlands.* Applied and Environmental Microbiology, 2005. **71**(3): p. 1453-1461.

36. Lopman, B.A., et al., *The Vast and Varied Global Burden of Norovirus: Prospects for Prevention and Control.* PLoS Med, 2016. **13**(4): p. e1001999.

37. Hata, A. and R. Honda, *Potential Sensitivity of Wastewater Monitoring for SARS-CoV-2: Comparison with Norovirus Cases.* Environmental science & technology, 2020. **54**(11): p. 6451-6452.

38. Olsen, J.S., et al., *Alternative Routes for Dissemination of Legionella pneumophila Causing Three Outbreaks in Norway.* Environmental Science & Technology, 2010. **44**(22): p. 8712-8717.

39. RIVM. *Legionella.* 2020 [cited 2020 08-12-2020 | 12:08].

40. Efstratiou, A., J.E. Ongerth, and P. Karanis, *Waterborne transmission of protozoan parasites: review of worldwide outbreaks-an update 2011–2016.* Water research, 2017. **114**: p. 14-22.

41. dos Santos Toledo, R., F.D.C. Martins, and R.L. Freire, *Waterborne Giardia and Cryptosporidium: contamination of human drinking water by sewage and cattle feces.* Semina: Ciências Agrárias, 2017. **38**(5): p. 3395-3415.

42. Huang, C., et al., *Environmental transport of emerging human-pathogenic Cryptosporidium species and subtypes through combined sewer overflow and wastewater.* Applied and environmental microbiology, 2017. **83**(16).

43. Widerström, M., et al., *Large outbreak of Cryptosporidium hominis infection transmitted through the public water supply, Sweden.* Emerging infectious diseases, 2014. **20**(4): p. 581-589.

44.   Ottoson, J., et al., *Removal of viruses, parasitic protozoa and microbial indicators in conventional and membrane processes in a wastewater pilot plant.* Water Research, 2006a. **40**(7): p. 1449-1457.

45.   Ottoson, J., et al., *Removal of noro-and enteroviruses, Giardia cysts, Cryptosporidium oocysts, and fecal indicators at four secondary wastewater treatment plants in Sweden.* Water Environment Research, 2006b. **78**(8): p. 828-834.

46.   Wen, Q., et al., *Fate of pathogenic microorganisms and indicators in secondary activated sludge wastewater treatment plants.* Journal of Environmental Management, 2009. **90**(3): p. 1442-1447.

47.   Odonkor, S.T. and J.K. Ampofo, *Escherichia coli as an indicator of bacteriological quality of water: an overview.* Microbiology research, 2013. **4**(1).

48.   Lee, S., M. Suwa, and H. Shigemura, *Occurrence and reduction of F-specific RNA bacteriophage genotypes as indicators of human norovirus at a wastewater treatment plant.* Journal of Water and Health, 2018.

49.   EPA, U., *Guidelines for water reuse EPA.* 2004.

50.   Environment_Agency_UK, *How to comply with your environmental permit. Additional guidance for Water Discharge and Groundwater (from point source) Activity Permits (EPR 7.01).* 2012.

51.   Madigan, M.T., J.M. Martinko, and J. Parker, *Brock biology of microorganisms.* Vol. 514. 1997: prentice hall Upper Saddle River, NJ.

52.   Tallon, P., et al., *Microbial indicators of faecal contamination in water: a current perspective.* Water, air, and soil pollution, 2005. **166**(1): p. 139-166.

53.   Bartram, J., R. Ballance, and W.H. Organization, *Water quality monitoring: a practical guide to the design and implementation of freshwater quality studies and monitoring programs.* 1996.

54.   Manero, A. and A.R. Blanch, *Identification of Enterococcus spp. with a biochemical key.* Applied and environmental microbiology, 1999. **65**(10): p. 4425-4430.

55.   De Luca, G., et al., *Removal of indicator bacteriophages from municipal wastewater by a full-scale membrane bioreactor and a conventional activated sludge process: Implications to water reuse.* Bioresource Technology, 2013. **129**: p. 526-531.

56.   Skraber, S., et al., *Survival of infectious Poliovirus-1 in river water compared to the persistence of somatic coliphages, thermotolerant coliforms and Poliovirus-1 genome.* Water Research, 2004. **38**(12): p. 2927-2933.

57.   Chahal, C., et al., *Pathogen and Particle Associations in Wastewater: Significance and Implications for Treatment and Disinfection Processes.* Adv Appl Microbiol, 2016. **97**: p. 63-119.

58. McMinn, B.R., N.J. Ashbolt, and A. Korajkic, *Bacteriophages as indicators of faecal pollution and enteric virus removal.* Letters in Applied Microbiology, 2017. **65**(1): p. 11-26.

59. Bijkerk P., et al., *State of infectious disease in the Netherlands 2015.* 2015, National Institute for Public Health and the Environment: The Netherlands. p. 76.

60. Lucena, F., et al., *Reduction of bacterial indicators and bacteriophages infecting faecal bacteria in primary and secondary wastewater treatments.* Journal of Applied Microbiology, 2004. **97**(5): p. 1069-1076.

61. Dias, E., J. Ebdon, and H. Taylor, *The application of bacteriophages as novel indicators of viral pathogens in wastewater treatment systems.* Water Research, 2018. **129**: p. 172-179.

62. Ma, J., et al., *Human infective potential of Cryptosporidium spp., Giardia duodenalis and Enterocytozoon bieneusi in urban wastewater treatment plant effluents.* Journal of Water and Health, 2016. **14**(3): p. 411-423.

63. van Beek, J., et al., *Comparison of norovirus genogroup I, II, and IV seroprevalence among children in the Netherlands, 1963, 1983, and 2006.* Journal of General Virology, 2016.

64. Liu, L., G. Hall, and P. Champagne, *Disinfection processes and mechanisms in wastewater stabilization ponds: a review.* Environmental Reviews, 2018(999): p. 1-13.

65. Carré, E., et al., *Impact of suspended particles on UV disinfection of activated-sludge effluent with the aim of reclamation.* Journal of Water Process Engineering, 2018. **22**: p. 87-93.

66. Amarasiri, M., et al., *Bacteriophage removal efficiency as a validation and operational monitoring tool for virus reduction in wastewater reclamation: review.* Water Research, 2017. **121**: p. 258-269.

67. Mandilara, G.D., et al., *Correlation between bacterial indicators and bacteriophages in sewage and sludge.* FEMS Microbiology Letters, 2006. **263**(1): p. 119-126.

68. Grabow, W., *Bacteriophages: update on application as models for viruses in water.* Water SA, 2001. **27**(2): p. 251-268.

69. Fauvel, B., et al., *Interactions of infectious F-specific RNA bacteriophages with suspended matter and sediment: towards an understanding of FRNAPH distribution in a river water system.* Science of the Total Environment, 2017. **574**: p. 960-968.

70. Roeleveld, P., J. Roorda, and M. Schaafsma, *The Dutch Roadmap for the WWTP of 2030.* 2010, STOWA report: Amersfoort, The Netherlands.

71. Matthews, B., et al., *Pathogen detection methodologies for wastewater and reservoirs*, in *Urban water security research alliance technical report.* 2010.

72.   Chahal, C., et al., *Pathogen and particle associations in wastewater: significance and implications for treatment and disinfection processes.* Advances in Applied Microbiology, 2016. **97**: p. 63-119.

73.   Guzman, C., et al., *Occurrence and levels of indicators and selected pathogens in different sludges and biosolids.* Journal of Applied Microbiology, 2007. **103**(6): p. 2420-2429.

74.   Mallory, L., et al., *Alternative prey: a mechanism for elimination of bacterial species by protozoa.* Applied and Environmental Microbiology, 1983. **46**(5): p. 1073-1079.

75.   Pronk, M., et al., *Full scale performance of the aerobic granular sludge process for sewage treatment.* Water Research, 2015. **84**: p. 207-217.

76.   APHA, *Standard Methods for the Examination of Water and Wastewater.* American Public Health Association/American Water Works Association/Water Environment Federation, Washington DC, USA, 2012.

77.   Anon, *ISO 10705-1*, in *Water Quality -Detection and Enumeration of Bacteriophages - Part 1: Enumeration of F-specific RNA Bacteriophages.* 2000, International Organization for Standarisation Geneva, Switzerland.

78.   Anon, *ISO 9308-1:2000*, in *Water Quality- Detection and Enumeration of Escherichea coli and Coliform Bacteria- Part 2: Membrane Filtration Method.* 2000, International Organization for Standarisation Geneva, Switzerland.

79.   Anon, *ISO 7899-2:2000*, in *Water Quality- Detection and Enumeration of Intestinal Enterococci- Part 2: Membrane Filtration Method.* 2000: International Organization for Standarisation, Geneva, Switzerland.

80.   R_Core_Team, *R: A language and environment for statistical computing. R Foundation for statistical computing, Vienna, Austria.* 2020.

81.   Pronk, M., et al., *Aerobic granular biomass technology: advancements in design, applications and further developments.* Water Practice and Technology, 2017. **12**(4): p. 987-996.

82.   Purnell, S., et al., *Removal of phages and viral pathogens in a full-scale MBR: Implications for wastewater reuse and potable water.* Water Research, 2016. **100**: p. 20-27.

83.   Stefanakis, A., et al., *Presence of bacteria and bacteriophages in full-scale trickling filters and an aerated constructed wetland.* Science of the Total Environment, 2019. **659**: p. 1135-1145.

84.   Zhang, K. and K. Farahbakhsh, *Removal of native coliphages and coliform bacteria from municipal wastewater by various wastewater treatment processes: implications to water reuse.* Water Research, 2007. **41**(12): p. 2816-2824.

85.   Thwaites, B.J., et al., *Comparing the performance of aerobic granular sludge versus conventional activated sludge for microbial log removal and effluent quality: implications for water reuse.* Water Research, 2018. **145**: p. 442-452.

86.   Hata, A., M. Kitajima, and H. Katayama, *Occurrence and reduction of human viruses, F-specific RNA coliphage genogroups and microbial indicators at a full-scale wastewater treatment plant in Japan.* Journal of Applied Microbiology, 2013. **114**(2): p. 545-554.

87.   Ulbricht, K., et al., *A mass balance approach to the fate of viruses in a municipal wastewater treatment plant during summer and winter seasons.* Water Science and Technology, 2013. **69**(2): p. 364-370.

88.   Madoni, P., *Microfauna biomass in activated sludge and biofilm.* Water Science and Technology, 1994. **29**(7): p. 63-66.

89.   Madoni, P., *Estimates of ciliated protozoa biomass in activated sludge and biofilm.* Bioresource Technology, 1994. **48**(3): p. 245-249.

90.   Young, R., *Bacteriophage lysis: mechanism and regulation.* Microbiological Reviews, 1992. **56**(3): p. 430-481.

91.   Hendricks, D.W., F.J. Post, and D.R. Khairnar, *Adsorption of bacteria on soils: experiments, thermodynamic rationale, and application.* Water, Air, and Soil Pollution, 1979. **12**(2): p. 219-232.

92.   Stevik, T.K., et al., *Retention and removal of pathogenic bacteria in wastewater percolating through porous media: a review.* Water Research, 2004. **38**(6): p. 1355-1367.

93.   Li, J., et al., *Aerobic granules dwelling vorticella and rotifers in an SBR fed with domestic wastewater.* Separation and Purification Technology, 2013. **110**: p. 127-131.

94.   Weber, S., et al., *Microbial composition and structure of aerobic granular sewage biofilms.* Applied and Environmental Microbiology, 2007. **73**(19): p. 6233-6240.

95.   Rabinovitch, A., I. Aviram, and A. Zaritsky, *Bacterial debris—an ecological mechanism for coexistence of bacteria and their viruses.* Journal of Theoretical Biology, 2003. **224**(3): p. 377-383.

96.   van Dijk, E.J., M. Pronk, and M.C.M. van Loosdrecht, *Controlling effluent suspended solids in the aerobic granular sludge process.* Water Research, 2018. **147**: p. 50-59.

97.   Bengtsson, S., et al., *Treatment of municipal wastewater with aerobic granular sludge.* Critical Reviews in Environmental Science and Technology, 2018. **48**(2): p. 119-166.

98.   Barrios-Hernández, M.L., et al., *Removal of bacterial and viral indicator organisms in full-scale aerobic granular sludge and conventional activated sludge systems.* Water Research X, 2020. **6**: p. 100040.

99.   van der Drift, C., et al., *Removal of Escherichia coli in wastewater by activated sludge.* Applied and environmental microbiology, 1977. **34**(3): p. 315-319.

100.   Curds, C.R.J.A.R.i.M., *The ecology and role of protozoa in aerobic sewage treatment processes.* 1982. **36**(1): p. 27-28.

101. Curds, C.R.J.A.Z., *The role of protozoa in the activated-sludge process.* 1973. **13**(1): p. 161-169.

102. Bales, R.C., et al., *MS-2 and poliovirus transport in porous media: Hydrophobic effects and chemical perturbations.* 1993. **29**(4): p. 957-963.

103. González, J.M. and C.A. Suttle, *Grazing by marine nanoflagellates on viruses and virus-sized particles: ingestion and digestion.* Marine Ecology Progress Series, 1993: p. 1-10.

104. Deng, L., et al., *Grazing of heterotrophic flagellates on viruses is driven by feeding behaviour.* Environmental Microbiology Reports, 2014. **6**(4): p. 325-330.

105. Dias, D., et al., *A review of bacterial indicator disinfection mechanisms in waste stabilisation ponds.* 2017. **16**(3): p. 517-539.

106. Vishniac, W. and M. Santer, *The thiobacilli.* Bacteriological reviews, 1957. **21**(3): p. 195.

107. Scoullos, I.M., et al., *Inactivation of indicator organisms on different surfaces after urban floods.* Science of The Total Environment, 2019: p. 135456.

108. Amaral, A.L., et al., *Use of chemometric analyses to assess biological wastewater treatment plants by protozoa and metazoa monitoring.* Environmental Monitoring and Assessment, 2018. **190**(9): p. 497.

109. Simonin, J.-P.J.C.E.J., *On the comparison of pseudo-first order and pseudo-second order rate laws in the modeling of adsorption kinetics.* 2016. **300**: p. 254-263.

110. Chao, A., *Nonparametric estimation of the number of classes in a population.* Scandinavian Journal of statistics, 1984: p. 265-270.

111. Kim, B.-R., et al., *Deciphering diversity indices for a better understanding of microbial communities.* J Microbiol Biotechnol, 2017. **27**(12): p. 2089-2093.

112. Madoni, P., G. Esteban, and G. Gorbi, *Acute toxicity of cadmium, copper, mercury, and zinc to ciliates from activated sludge plants.* Bulletin of environmental contamination and toxicology, 1992. **49**(6): p. 900-905.

113. De Kreuk, M. and M. Van Loosdrecht, *Selection of slow growing organisms as a means for improving aerobic granular sludge stability.* Water Science & Technology, 2004. **49**(11): p. 9-17.

114. Adav, S.S., et al., *Aerobic granular sludge: recent advances.* Biotechnol Adv, 2008. **26**(5): p. 411-23.

115. Winkler, M.-K., et al., *Selective sludge removal in a segregated aerobic granular biomass system as a strategy to control PAO–GAO competition at high temperatures.* Water research, 2011. **45**(11): p. 3291-3299.

116. Sato, C., S.W. Leung, and J.L. Schnoor, *Toxic response of Nitrosomonas europaea to copper in inorganic medium and wastewater.* Water Research, 1988. **22**(9): p. 1117-1127.

117.   Lee, N.M. and T. Welander, *Influence of Predators on Nitrification in Aerobic Biofilm Processes.* Water Science and Technology, 1994. **29**(7): p. 355-363.

118.   Chong, S., et al., *The performance enhancements of upflow anaerobic sludge blanket (UASB) reactors for domestic sludge treatment – A State-of-the-art review.* Water Research, 2012. **46**(11): p. 3434-3470.

119.   Vymazal, J., *Removal of enteric bacteria in constructed treatment wetlands with emergent macrophytes: a review.* Journal of Environmental Science and Health, 2005. **40**(6-7): p. 1355-1367.

120.   Corpuz, M.V.A., et al., *Viruses in wastewater: occurrence, abundance and detection methods.* Science of The Total Environment, 2020. **745**: p. 140910.

121.   Martín-Díaz, J., et al., *Review: Indicator bacteriophages in sludge, biosolids, sediments and soils.* Environmental Research, 2020. **182**: p. 109133.

122.   Zeng, Q., et al., *Electrochemical pretreatment for stabilization of waste activated sludge: Simultaneously enhancing dewaterability, inactivating pathogens and mitigating hydrogen sulfide.* Water Research, 2019. **166**: p. 115035.

123.   Zhang, L., et al., *Improving heavy metals removal, dewaterability and pathogen removal of waste activated sludge using enhanced chemical leaching.* Journal of Cleaner Production, 2020. **271**: p. 122512.

124.   Lochmatter, S., G. Gonzalez-Gil, and C. Holliger, *Optimized aeration strategies for nitrogen and phosphorus removal with aerobic granular sludge.* Water Research, 2013. **47**(16): p. 6187-6197.

125.   Van Loosdrecht, M., et al., *Biofilm structures.* Water Science and Technology, 1995. **32**(8): p. 35.

126.   Fenchel, T., *Protozoa and Oxygen.* Acta protozoologica, 2014. **53**: p. 3-12.

127.   McKinney, R.E. and A. Gram, *Protozoa and activated sludge.* Sewage and Industrial Wastes, 1956. **28**(10): p. 1219-1231.

128.   Madoni, P., *Protozoa in wastewater treatment processes: a minireview.* Italian Journal of Zoology, 2011. **78**(1): p. 3-11.

129.   Nisbet, B., *Nutrition and feeding strategies in protozoa.* 1984: Croom Helm, London.

130.   de Kreuk, M., et al., *Behavior of polymeric substrates in an aerobic granular sludge system.* Water Research, 2010. **44**(20): p. 5929-5938.

131.   Barrios-Hernández, M.L., et al., *Effect of the co-treatment of synthetic faecal sludge and wastewater in an aerobic granular sludge system.* Science of The Total Environment, 2020: p. 140480.

132.   Varma, M.M., H.E. Finley, and G.H.J.J. Bennett, *Population dynamics of protozoa in wastewater.* 1975: p. 85-92.

133. Dubber, D. and N.J.W.r. Gray, *The effect of anoxia and anaerobia on ciliate community in biological nutrient removal systems using laboratory-scale sequencing batch reactors (SBRs).* 2011. **45**(6): p. 2213-2226.

134. Gao, F., et al., *The all-data-based evolutionary hypothesis of ciliated protists with a revised classification of the phylum Ciliophora (Eukaryota, Alveolata).* Scientific Reports, 2016. **6**: p. 24874.

135. Leal, C., A.L. Amaral, and M. de Lourdes Costa, *Microbial-based evaluation of foaming events in full-scale wastewater treatment plants by microscopy survey and quantitative image analysis.* Environmental Science and Pollution Research, 2016. **23**(15): p. 15638-15650.

136. Adl, S.M., et al., *Revisions to the classification, nomenclature, and diversity of eukaryotes.* Journal of Eukaryotic Microbiology, 2019. **66**(1): p. 4-119.

137. Belar, K., *Untersuchungen über Thecamöben der Chlamydophrys-Gruppe.* Arch. Protistenkd, 1921. **43**: p. 287-354.

138. Öztoprak, H., et al., *What Drives the Diversity of the Most Abundant Terrestrial Cercozoan Family (Rhogostomidae, Cercozoa, Rhizaria)?* Microorganisms, 2020. **8**(8): p. 1123.

139. Howe, A.T., et al., *Novel Cultured Protists Identify Deep-branching Environmental DNA Clades of Cercozoa: New Genera Tremula, Micrometopion, Minimassisteria, Nudifila, Peregrinia.* Protist, 2011. **162**(2): p. 332-372.

140. Dumack, K., et al., *Rhogostomidae (Cercozoa) from soils, roots and plant leaves (Arabidopsis thaliana): Description of Rhogostoma epiphylla sp. nov. and R. cylindrica sp. nov.* European Journal of Protistology, 2017. **60**: p. 76-86.

141. Berman, J.J., *Chapter 20 - Ciliophora (Ciliates)*, in *Taxonomic Guide to Infectious Diseases*, J.J. Berman, Editor. 2012, Academic Press: Boston. p. 111-112.

142. Lynn, D., *The ciliated protozoa: characterization, classification, and guide to the literature.* 2008: Springer Science & Business Media.

143. Hirakata, Y., et al., *Effects of predation by protists on prokaryotic community function, structure, and diversity in anaerobic granular sludge.* Microbes and environments, 2016: p. ME16067.

144. Schulz, S., S. Wagener, and N. Pfennig, *Utilization of various chemotrophic and phototrophic bacteria as food by the anaerobic ciliate Trimyema compressum.* European Journal of Protistology, 1990. **26**(2): p. 122-131.

145. Matsunaga, K., K. Kubota, and H. Harada, *Molecular diversity of eukaryotes in municipal wastewater treatment processes as revealed by 18S rRNA gene analysis.* Microbes and environments, 2014. **29**(4): p. 401-407.

146. Saha, D. and R. Mukherjee, *Ameliorating the antimicrobial resistance crisis: phage therapy.* IUBMB Life, 2019. **71**(7): p. 781-790.

147. Salas, M. and M. de Vega, *Replication of Bacterial Viruses*, in *Encyclopedia of Virology (Third Edition)*, B.W.J. Mahy and M.H.V. Van Regenmortel, Editors. 2008, Academic Press: Oxford. p. 399-406.

148. Berzin, V., G. Rosenthal, and E.J. Gren, *Cellular macromolecule synthesis in Escherichia coli infected with bacteriophage MS2*. Eur J Biochem, 1974. **45**(1): p. 233-42.

149. Zhang, Y., H.K. Hunt, and Z. Hu, *Application of bacteriophages to selectively remove Pseudomonas aeruginosa in water and wastewater filtration systems*. Water Research, 2013. **47**(13): p. 4507-4518.

150. Marks, T. and R. Sharp, *Bacteriophages and biotechnology: a review*. Journal of Chemical Technology & Biotechnology: International Research in Process, Environmental & Clean Technology, 2000. **75**(1): p. 6-17.

151. Khan, M.A., et al., *Bacteriophages isolated from activated sludge processes and their polyvalency*. Water research, 2002. **36**(13): p. 3364-3370.

152. Hantula, J., et al., *Ecology of bacteriophages infecting activated sludge bacteria*. Appl. Environ. Microbiol., 1991. **57**(8): p. 2147-2151.

153. Kuzmanovic, D.A., et al., *Bacteriophage MS2: Molecular Weight and Spatial Distribution of the Protein and RNA Components by Small-Angle Neutron Scattering and Virus Counting*. Structure, 2003. **11**(11): p. 1339-1348.

154. Jürgens, K., *Predation on Bacteria and Bacterial Resistance Mechanisms: Comparative AspectsAmong Different Predator Groups in Aquatic Systems*, in *Predatory Prokaryotes: Biology, Ecology and Evolution*, E. Jurkevitch, Editor. 2007, Springer Berlin Heidelberg: Berlin, Heidelberg. p. 57-92.

155. Beun, J., M. Van Loosdrecht, and J. Heijnen, *Aerobic granulation in a sequencing batch airlift reactor*. Water Research, 2002. **36**(3): p. 702-712.

156. Goberna, M., et al., *Prokaryotic communities and potential pathogens in sewage sludge: Response to wastewaster origin, loading rate and treatment technology*. Science of The Total Environment, 2018. **615**: p. 360-368.

157. van Dijk, E.J.H., M. Pronk, and M.C.M. van Loosdrecht, *A settling model for full-scale aerobic granular sludge*. Water Research, 2020. **186**: p. 116135.

158. Pratt, L.A. and R.J.M.m. Kolter, *Genetic analysis of Escherichia coli biofilm formation: roles of flagella, motility, chemotaxis and type I pili*. 1998. **30**(2): p. 285-293.

159. Tay, J.-H., H.-L. Xu, and K.-C.J.J.o.E.E. Teo, *Molecular mechanism of granulation. I: H+ translocation-dehydration theory*. 2000. **126**(5): p. 403-410.

160. Tay, J.H., Q.S. Liu, and Y.J.J.o.A.M. Liu, *Microscopic observation of aerobic granulation in sequential aerobic sludge blanket reactor*. 2001. **91**(1): p. 168-175.

161. Liu, Y., et al., *The role of cell hydrophobicity in the formation of aerobic granules*. 2003. **46**(4): p. 0270-0274.

162. Pallares-Vega, R., et al., *Annual dynamics of antimicrobials and resistance determinants in flocculent and aerobic granular sludge treatment systems.* Water Research, 2020: p. 116752.

163. Hirakata, Y., et al., *Temporal variation of eukaryotic community structures in UASB reactor treating domestic sewage as revealed by 18S rRNA gene sequencing.* Scientific Reports, 2019. **9**(1): p. 12783.

164. Ingallinella, A., et al., *The challenge of faecal sludge management in urban areas-strategies, regulations and treatment options.* Water Science and Technology, 2002. **46**(10): p. 285-294.

165. Strande, L., M. Ronteltap, and D. Brdjanovic, *Faecal Sludge Management: Systems Approach for Implementation and Operation.* 2014: IWA Publishing.

166. Strauss, M. and A. Montangero, *FS management–review of practices, problems and initiatives.* 2002, EAWAG/SANDEC publications: Dubendorf.

167. Lopez-Vazquez, C.M., et al., *Co-treatment of faecal sludge in municipal wastewater treatment plants*, in *Faecal Sludge Management —Systems Approach Implementation and Operation.*, L. Strande, M. Ronteltap, and D. Brdjanovic, Editors. 2014, IWA Publishing: London, UK. p. 177-198.

168. Tayler, K., *Faecal Sludge and Septage Treatment: A Guide for Low and Middle Income Countries.* 2018: Practical Action Publishing.

169. Siegrist, R.L., *Treatment Using Septic Tanks*, in *Decentralized Water Reclamation Engineering: A Curriculum Workbook.* 2017, Springer International Publishing. p. 237-288.

170. Strauss, M., et al., *Treating faecal sludges in ponds.* Water Science and Technology, 2000. **42**(10-11): p. 283-290.

171. Heinss, U. and M. Strauss, *Co-treatment of faecal sludge and wastewater in tropical climates.* SOS-Management of sludges from on-site sanitation. EAWAG/SANDEC., 1999.

172. Penn, R., et al., *Review of synthetic human faeces and faecal sludge for sanitation and wastewater research.* Water Research, 2018. **132**: p. 222-240.

173. Udert, K.M. and M. Wächter, *Complete nutrient recovery from source-separated urine by nitrification and distillation.* Water Research, 2012. **46**(2): p. 453-464.

174. Winkler, M.K.H., et al., *Unravelling the reasons for disproportion in the ratio of AOB and NOB in aerobic granular sludge.* Applied Microbiology and Biotechnology, 2012. **94**(6): p. 1657-1666.

175. Rasband, W.S., *ImageJ, U. S. National Institutes of Health, Bethesda, Maryland, USA, https://imagej.nih.gov/ij/.* 1997-2018.

176. Bower, K.M., *Water supply and sanitation of Costa Rica.* Environmental Earth Sciences, 2014. **71**(1): p. 107-123.

177. Marui, J., et al., *Reduction of the degradation activity of umami-enhancing purinic ribonucleotide supplement in miso by the targeted suppression of acid*

phosphatases in the Aspergillus oryzae starter culture. International Journal of Food Microbiology, 2013. **166**(2): p. 238-243.

178. Methven, L., *Natural food and beverage flavour enhancer*, in *Natural Food Additives, Ingredients and Flavourings*, D. Baines and R. Seal, Editors. 2012, Woodhead Publishing: Cambridge. p. 76-99.

179. Nout, R., *18 - Quality, safety, biofunctionality and fermentation control in soya*, in *Advances in Fermented Foods and Beverages*, W. Holzapfel, Editor. 2015, Woodhead Publishing. p. 409-434.

180. Schuler, A.J. and D. Jenkins, *Enhanced biological phosphorus removal from wastewater by biomass with different phosphorus contents, part I: experimental results and comparison with metabolic models.* Water Environment Research, 2003. **75**(6): p. 485-498.

181. Bassin, J.P., et al., *Improved phosphate removal by selective sludge discharge in aerobic granular sludge reactors.* Biotechnology and Bioengineering, 2012. **109**(8): p. 1919-1928.

182. Lopez-Vazquez, C.M., et al., *Modeling the PAO–GAO competition: effects of carbon source, pH and temperature.* Water Research, 2009. **43**(2): p. 450-462.

183. Bassin, J.P., et al., *Measuring biomass specific ammonium, nitrite and phosphate uptake rates in aerobic granular sludge.* Chemosphere, 2012. **89**(10): p. 1161-1168.

184. Bassin, J.P., et al., *Ammonium adsorption in aerobic granular sludge, activated sludge and anammox granules.* Water Research, 2011. **45**(16): p. 5257-5265.

185. Morgenroth, E., R. Kommedal, and P. Harremoës, *Processes and modeling of hydrolysis of particulate organic matter in aerobic wastewater treatment–a review.* Water Science and Technology, 2002. **45**(6): p. 25-40.

186. Saito, T., D.v. Brdjanovic, and M. Van Loosdrecht, *Effect of nitrite on phosphate uptake by phosphate accumulating organisms.* Water Research, 2004. **38**(17): p. 3760-3768.

187. Szabó, E., et al., *Effects of wash-out dynamics on nitrifying bacteria in aerobic granular sludge during start-up at gradually decreased settling time.* Water, 2016. **8**(5): p. 172.

188. Mosquera-Corral, A., et al., *Effects of oxygen concentration on N-removal in an aerobic granular sludge reactor.* Water Research, 2005. **39**(12): p. 2676-86.

189. Rocktäschel, T., et al., *Influence of the granulation grade on the concentration of suspended solids in the effluent of a pilot scale sequencing batch reactor operated with aerobic granular sludge.* Separation and Purification Technology, 2015. **142**: p. 234-241.

190. Khan, A.A., M. Ahmad, and A. Giesen, *NEREDA®: an emerging technology for sewage treatment.* Water Practice and Technology, 2015. **10**(4): p. 799-805.

191. Corsino, S.F., et al., *Aerobic granular sludge treating high strength citrus wastewater: Analysis of pH and organic loading rate effect on kinetics, performance and stability.* Journal of Environmental Management, 2018. **214**: p. 23-35.

192. Wagner, J., et al., *Effect of particulate organic substrate on aerobic granulation and operating conditions of sequencing batch reactors.* Water Research, 2015. **85**: p. 158-166.

193. Cetin, E., et al., *Effects of high-concentration influent suspended solids on aerobic granulation in pilot-scale sequencing batch reactors treating real domestic wastewater.* Water Research, 2018. **131**: p. 74-89.

194. Zhou, J.-h., et al., *Optimizing granules size distribution for aerobic granular sludge stability: Effect of a novel funnel-shaped internals on hydraulic shear stress.* Bioresource Technology, 2016. **216**: p. 562-570.

195. Liu, Y.-Q. and J.-H. Tay, *The competition between flocculent sludge and aerobic granules during the long-term operation period of granular sludge sequencing batch reactor.* Environmental Technology, 2012. **33**(23): p. 2619-2626.

196. Zhang, B., et al., *Microbial population dynamics during sludge granulation in an anaerobic–aerobic biological phosphorus removal system.* Bioresource Technology, 2011. **102**(3): p. 2474-2480.

197. Araújo dos Santos, L., et al., *Relationship between protozoan and metazoan communities and operation and performance parameters in a textile sewage activated sludge system.* European Journal of Protistology, 2014. **50**(4): p. 319-328.

198. Fiałkowska, E. and A. Pajdak-Stós, *The role of Lecane rotifers in activated sludge bulking control.* Water Research, 2008. **42**(10): p. 2483-2490.

199. Assress, H.A., et al., *Diversity, Co-occurrence and Implications of Fungal Communities in Wastewater Treatment Plants.* Scientific Reports, 2019. **9**(1): p. 14056.

200. Ye, J., et al., *Biosorption of chromium from aqueous solution and electroplating wastewater using mixture of Candida lipolytica and dewatered sewage sludge.* Bioresource Technology, 2010. **101**(11): p. 3893-3902.

201. Siqueira-Castro, I.C.V., et al., *First report of predation of Giardia sp. cysts by ciliated protozoa and confirmation of predation of Cryptosporidium spp. oocysts by ciliate species.* Environmental Science and Pollution Research, 2016. **23**(11): p. 11357-11362.

202. Stott, R., et al., *Protozoan predation as a mechanism for the removal of cryptosporidium oocysts from wastewaters in constructed wetlands.* Water Science and Technology, 2001. **44**(11-12): p. 191-198.

203. Burki, F., et al., *The New Tree of Eukaryotes.* Trends in Ecology & Evolution, 2020. **35**(1): p. 43-55.

204. Paziewska-Harris, A., G. Schoone, and H.D.F.H. Schallig, *An easy 'one tube' method to estimate viability of Cryptosporidium oocysts using real-time qPCR.* Parasitology Research, 2016. **115**(7): p. 2873-2877.

205. Small, E.B. and D.S.J.S. Marszalek, *Scanning electron microscopy of fixed, frozen, and dried protozoa.* 1969. **163**(3871): p. 1064-1065.

206. Martín-Cereceda, M., et al., *Classification of the peritrich ciliate Opisthonecta matiensis (Martín-Cereceda et al. 1999) as Telotrochidium matiense nov. comb., based on new observations and SSU rDNA phylogeny.* European Journal of Protistology, 2007. **43**(4): p. 265-279.

207. Sun, P., et al., *Vorticella Linnaeus, 1767 (Ciliophora, Oligohymenophora, Peritrichia) is a grade not a clade: redefinition of Vorticella and the families Vorticellidae and Astylozoidae using molecular characters derived from the gene coding for small subunit ribosomal RNA.* Protist, 2012. **163**(1): p. 129-42.

208. Chouari, R., et al., *Eukaryotic molecular diversity at different steps of the wastewater treatment plant process reveals more phylogenetic novel lineages.* World Journal of Microbiology and Biotechnology, 2017. **33**(3): p. 44.

209. Xia, J., et al., *Microbial community structure and function in aerobic granular sludge.* Applied Microbiology and Biotechnology, 2018. **102**(9): p. 3967-3979.

210. Caporaso, J.G., et al., *QIIME allows analysis of high-throughput community sequencing data.* Nature methods, 2010. **7**(5): p. 335-336.

211. Edgar, R.C., *UPARSE: highly accurate OTU sequences from microbial amplicon reads.* Nat Methods, 2013. **10**(10): p. 996-8.

212. Shade, A. and J. Handelsman, *Beyond the Venn diagram: the hunt for a core microbiome.* Environmental Microbiology, 2012. **14**(1): p. 4-12.

213. Schloss, P.D., et al., *Introducing mothur: Open-Source, Platform-Independent, Community-Supported Software for Describing and Comparing Microbial Communities.* Applied and Environmental Microbiology, 2009. **75**(23): p. 7537-7541.

214. Quast, C., et al., *The SILVA ribosomal RNA gene database project: improved data processing and web-based tools.* Nucleic acids research, 2012. **41**(D1): p. D590-D596.

215. Segata, N., et al., *Metagenomic biomarker discovery and explanation.* Genome Biology, 2011. **12**(6): p. R60.

216. Lee, S.-H., H.-J. Kang, and H.-D. Park, *Influence of influent wastewater communities on temporal variation of activated sludge communities.* Water Research, 2015. **73**: p. 132-144.

217. Winkler, M.-K.H., et al., *Microbial diversity differences within aerobic granular sludge and activated sludge flocs.* Applied Microbiology and Biotechnology, 2013. **97**(16): p. 7447-7458.

218. Fenchel, T., *Ecology of Protozoa: The biology of free-living phagotropic protists.* 2013: Springer-Verlag.

219. Pauli, W., K. Jax, and S. Berger, *Protozoa in wastewater treatment: function and importance*, in *Biodegradation and Persistance.* 2001, Springer. p. 203-252.

220. Ntougias, S., S. Tanasidis, and P. Melidis, *Microfaunal indicators, Ciliophora phylogeny and protozoan population shifts in an intermittently aerated and fed bioreactor.* Journal of hazardous materials, 2011. **186**(2-3): p. 1862-1869.

221. Evans, T.N. and R.J. Seviour, *Estimating biodiversity of fungi in activated sludge communities using culture-independent methods.* Microbial ecology, 2012. **63**(4): p. 773-786.

222. Madoni, P., *A sludge biotic index (SBI) for the evaluation of the biological performance of activated sludge plants based on the microfauna analysis.* Water Research, 1994. **28**(1): p. 67-75.

223. Ryu, S., et al., *Vorticella: a protozoan for bio-inspired engineering.* Micromachines, 2017. **8**(1): p. 4.

224. Reid, R., *Fluctuations in populations of 3 Vorticella species from an activated-sludge sewage plant.* The Journal of Protozoology, 1969. **16**(1): p. 103-111.

225. Poole, J., *A study of the relationship between the mixed liquor fauna and plant performance for a variety of activated sludge sewage treatment works.* Water Research, 1984. **18**(3): p. 281-287.

226. Rehman, A., F.R. Shakoori, and A.R. Shakoori, *Resistance and uptake of heavy metals by Vorticella microstoma and its potential use in industrial wastewater treatment.* Environmental Progress & Sustainable Energy, 2010. **29**(4): p. 481-486.

227. Rehman, A., R. Farah, and A.R. Shakoori, *Potential use of a ciliate, Vorticella microstoma, surviving in lead containing industrial effluents, in waste water treatment.* Pakistan Journal of Zoology, 2007. **39**(4): p. 259.

228. Bramucci, M.G. and V. Nagarajan, *Inhibition of Vorticella microstoma stalk formation by wheat germ agglutinin.* J Eukaryot Microbiol, 2004. **51**(4): p. 425-7.

229. Calvo, P., et al., *Ultrastructure, eneystment and cyst wall composition of the resting cyst of the peritrich ciliate Opisthonecta henneguyi.* Journal of Eukaryotic Microbiology, 2003. **50**(1): p. 49-56.

230. Papadimitriou, C.A., et al., *Investigation of protozoa as indicators of wastewater treatment efficiency in constructed wetlands.* Desalination, 2010. **250**(1): p. 378-382.

231. Madoni, P. and P.F. Ghetti, *The structure of Ciliated Protozoa communities in biological sewage-treatment plants.* Hydrobiologia, 1981. **83**(2): p. 207-215.

232. Evans, M.S., L.M. Sicko-Goad, and M. Omair, *Seasonal Occurrence of Tokophrya quadripartita (Suctoria) as Epibionts on Adult Limnocalanus*

macrurus (Copepoda: Calanoida) in Southeastern Lake Michigan. Transactions of the American Microscopical Society, 1979. **98**(1): p. 102-109.

233. Özer, A., *Trichodina domerguei Wallengren, 1897 (Ciliophora: Peritrichia) Infestations on the Round Goby, Neogobius melanostomus Pallas, 1811in Relation to Seasonality and Host Factors.* Comparative Parasitology, 2003. **70**(2): p. 132-135, 4.

234. Weber, S.D., et al., *The diversity of fungi in aerobic sewage granules assessed by 18S rRNA gene and ITS sequence analyses.* FEMS microbiology ecology, 2009. **68**(2): p. 246-254.

235. Sharaf, A., B. Guo, and Y. Liu, *Impact of the filamentous fungi overgrowth on the aerobic granular sludge process.* Bioresource Technology Reports, 2019. 7: p. 100272.

236. Liu, J., et al., *Analysis of bacterial, fungal and archaeal populations from a municipal wastewater treatment plant developing an innovative aerobic granular sludge process.* World J Microbiol Biotechnol, 2017. **33**(1): p. 14.

237. Zhang, H., et al., *Disentangling the Drivers of Diversity and Distribution of Fungal Community Composition in Wastewater Treatment Plants Across Spatial Scales.* Frontiers in Microbiology, 2018. **9**(1291).

238. Morales, D.K., et al., *Control of Candida albicans metabolism and biofilm formation by Pseudomonas aeruginosa phenazines.* MBio, 2013. **4**(1).

239. Takahashi, Y., et al., *Fibrophrys columna gen. nov., sp. nov: A member of the family Amphifilidae.* European Journal of Protistology, 2016. **56**: p. 41-50.

240. Harwood, V.J., et al., *Validity of the Indicator Organism Paradigm for Pathogen Reduction in Reclaimed Water and Public Health Protection.* Applied and Environmental Microbiology, 2005. **71**(6): p. 3163-3170.

241. McClung, R.P., et al., *Waterborne disease outbreaks associated with environmental and undetermined exposures to water—United States, 2013–2014.* MMWR. Morbidity and mortality weekly report, 2017. **66**(44): p. 1222.

242. Caicedo, C., et al., *Legionella occurrence in municipal and industrial wastewater treatment plants and risks of reclaimed wastewater reuse.* Water research, 2019. **149**: p. 21-34.

243. WHO. *WHO priority pathogens list for R&D of new antibiotics.* News release 2020 [cited 2020 13-08-2020].

244. Randazzo, W., et al., *SARS-CoV-2 RNA in wastewater anticipated COVID-19 occurrence in a low prevalence area.* Water Research, 2020: p. 115942.

245. Medema, G., et al., *Presence of SARS-Coronavirus-2 in sewage.* MedRxiv, 2020.

246. Wang, L., et al., *Recent advances on biosorption by aerobic granular sludge.* Journal of hazardous materials, 2018. **357**: p. 253-270.

247. Mamane, H., *Impact of particles on UV disinfection on water and wastewater effluents: a review.* Vol. 24. 2008. 67-157.

248. Siew Herng, C., et al., *Microbial Predation Accelerates Granulation and Modulates Microbial Community Composition.* BMC Microbiology, 2021.

249. Boamah, D.K., et al., *From Many Hosts, One Accidental Pathogen: The Diverse Protozoan Hosts of Legionella.* Frontiers in Cellular and Infection Microbiology, 2017. 7(477).

250. Scheikl, U., et al., *Free-living amoebae (FLA) co-occurring with legionellae in industrial waters.* European journal of protistology, 2014. **50**(4): p. 422-429.

251. McDougald, D. and S.R. Longford, *Protozoa hosts lead to virulence.* Nature Microbiology, 2020. **5**(4): p. 535-535.

LIST OF ACRONYMS

AOB	Ammonium-oxidising bacteria
AGS	Aerobic granular sludge
BOD	Biological oxygen demand
CAS	Conventional activated sludge
CDC	Centers for Disease Control and Prevention
COD	Chemical oxygen demand
CFU	Colony forming units
DNA	Deoxyribonucleic Acid
FS	Faecal sludge
FIOs	Faecal indicator organisms
GAO	Glycogen accumulating organisms
HRT	Hydraulic retention time
NOB	Nitrite-oxidising bacteria
OHO	Ordinary heterotrophic organism
OTUs	Operational Taxonomic Units
PAOs	Phosphate accumulating organism
PFU	Plaque-forming unit
RNA	Ribonucleic Acid
SBR	Sequencing batch reactor
SEM	Scanning electron microscopy
SVI	Sludge volume index
TSS	Total suspended solids
VSS	Volatile suspended solids
WWTP	Wastewater treatment plant

LIST OF TABLES

LIST OF FIGURES

ABOUT THE AUTHOR

Mary Luz BARRIOS HERNANDEZ

25-03-1986   Born in San José, Costa Rica

EDUCATION

2016 -2020   **PhD Environmental Biotechnology and Sanitary Engineering**

Technical University of Delft
IHE Delft, Institute for Water Education

Thesis: Pathogen removal in aerobic granular sludge treatment systems.

Promotors: Prof. M.C.M. van Loosdrecht, Prof. D. Brdjanovic, and C.M Hooijmans

2014 -2016   **Master of Science on Urban Water and Sanitation**

Sanitary Engineering

UNESCO-IHE Institute for Water Education

2006 -2011   **Environmental Engineering (Licenciatura Ingeniería Ambiental),**

Instituto Tecnológico de Costa Rica.

Journals publications

» **Barrios-Hernández, M.L.**, Bettinelli-Travián, C. Mora-Cabrera K., Vanegas-Camero M.C., Garcia, H.A., van de Vossenberg, Prats D., Brdjanovic, D. van Loosdrecht, M.C.M., and Hooijmans, C.M. Unravelling the *E. coli* and MS2 bacteriophage removal mechanisms in aerobic granular sludge systems.Water Research 195, 116992.

» **Barrios-Hernández, M.L.**, Buenaño-Vargas, C., García, H., Brdjanovic, D., van Loosdrecht, M.C.M. and Hooijmans, C.M. 2020. Effect of the co-treatment of synthetic faecal sludge and wastewater in an aerobic granular sludge system. Science of the Total Environment, 140480.

» **Barrios-Hernández, M.L.,** Pronk, M., Garcia, H., Boersma, A., Brdjanovic, D., van Loosdrecht, M.C.M. and Hooijmans, C.M. 2020. Removal of bacterial and viral indicator organisms in full-scale aerobic granular sludge and conventional activated sludge systems. Water Research X 6, 100040.

Conference proceedings

» Effect of the co-treatment of (synthetic) septic sludge and wastewater in an aerobic granular sludge system. Oral presentation. Presented at *PhD Symposium Innovations for sustainability,* 10th & 11th October 2019. IHE Delft.

» Pathogen removal mechanisms in aerobic granular sludge systems. Poster and flash presentation. *20th IWA HRWM Symposium,* September 2019, Vienna, Austria.

» Aerobic granular sludge systems. Presented at PhD Symposium 2018*: Nature for Water – Overcoming Water Challenges with Sustainable Solutions,* IHE Delft.

» Comparison of the fate of a pathogen indicator (*E. coli*) between a lab-scale Aerobic Granular Sludge and Activated Sludge process. Flash oral contribution and poster presentation. Presented at *IWA Biofilms: Granular Sludge System Conference*, 2018. Delft

» Fate of *E. coli* in Aerobic Granular Sludge and Activated Sludge laboratory wastewater treatment reactor. Poster session presented at *Symposium on Microbiological Methods for Waste and Wastewater Resource Recovery.* 2017. Delft

*Netherlands Research School for the
Socio-Economic and Natural Sciences of the Environment*

D I P L O M A

for specialised PhD training

The Netherlands research school for the
Socio-Economic and Natural Sciences of the Environment
(SENSE) declares that

Mary Luz
Barrios Hernández

born on 25 March 1986 in San José, Costa Rica

has successfully fulfilled all requirements of the
educational PhD programme of SENSE.

Delft, 30 September 2021

Chair of the SENSE board   The SENSE Director

Prof. dr. Martin Wassen   Prof. Philipp Pattberg

The SENSE Research School has been accredited by the Royal Netherlands Academy of Arts and Sciences (KNAW)

KONINKLIJKE NEDERLANDSE
AKADEMIE VAN WETENSCHAPPEN

The SENSE Research School declares that **Mary Luz Barrios Hernández** has successfully fulfilled all requirements of the educational PhD programme of SENSE with a work load of 43.8 EC, including the following activities:

SENSE PhD Courses

- Environmental research in context (2017)
- Research in context activity: 'Co-organizing Wastewater and Sanitation Workshop (San José, Costa Rica – 18-20 July 2017)'
- Principles of ecological and evolutionary Genomics (2018)
- Introduction to R for Statistical Analysis (2019)

Other PhD and Advanced MSc Courses

- How to become effective in a network conversation, TU-Delft (2017)
- Designing Scientific Posters and Theses with Adobe InDesign, TU-Delft (2017)
- Using creativity to maximize productivity and innovation, TU-Delft (2017)
- Coaching Individual Students and Project Groups, TU-Delft (2017)
- Molecular biology fundamentals training, PETAL (2018)
- Advanced course: Environmental Biotechnology, TU-Delft (2017)
- Career Development - Exploring a research career outside academia, TU-Delft (2019)
- Two-day scientific writing workshop, IHE Delft (2019)

Management and Didactic Skills Training

- Supervising four MSc student with thesis entitled (2017-2020)
- Co-organizing of 'Aerobic granular sludge, an advanced technology for the simultaneous organic matter and nutrient removal', Wastewater and Sanitation Workshop, 18-20 July 2017, San José, Costa Rica.

Oral Presentations

- *Comparison of the fate of a pathogen indicator (E.coli) between a lab-scale aerobic granular sludge and activated sludge process.* IWA Biofilms: Granular Sludge Conference, 18-21 March 2018, TU Delft, The Netherlands
- *Faecal indicator removal in full-scale aerobic granular sludge systems.* Nature for Water: Overcoming Water Challenges with Sustainable Solution, IHE-Delft, 1-2 October 2018, Delft, The Netherlands
- *Effect of the co-treatment of (synthetic) septic sludge and wastewater in an aerobic granular sludge system.* PhD Symposium Innovations for sustainability, 10-11 October 2019. IHE Delft, The Netherlands
- *Pathogen removal mechanisms in aerobic granular sludge systems.* 20th IWA HRWM Symposium, 15-20 September 2019, Vienna , Austria

SENSE coordinator PhD education

Dr. ir. Peter Vermeulen